NOKIA
SMARTPHONE
HACKS ™

Other resources from O'Reilly

Related titles

Palm and Treo Hacks

Talk Is Cheap

Treo Fan Book

PC Hardware Buyer's Guide

Don't Click on the Blue E!

PC Hardware Annoyances

Smart Home Hacks

Car PC Hacks

Building the Perfect PC

Make: Technology on Your Time

Wireless Hacks

Hacks Series Home

hacks.oreilly.com is a community site for developers and power users of all stripes. Readers learn from each other as they share their favorite tips and tools for Mac OS X, Linux, Google, Windows XP, and more.

oreilly.com

oreilly.com is more than a complete catalog of O'Reilly books. You'll also find links to news, events, articles, weblogs, sample chapters, and code examples.

oreillynet.com is the essential portal for developers interested in open and emerging technologies, including new platforms, programming languages, and operating systems.

Conferences

O'Reilly brings diverse innovators together to nurture the ideas that spark revolutionary industries. We specialize in documenting the latest tools and systems, translating the innovator's knowledge into useful skills for those in the trenches. Visit *conferences.oreilly.com* for our upcoming events.

Safari Books Online (*safaribooksonline.com*) is the premier online reference library for programmers and IT professionals. Conduct searches across more than 1,000 books. Subscribers can zero in on answers to time-critical questions in a matter of seconds. Read the books on your Bookshelf from cover to cover or simply flip to the page you need. Try it today for free.

NOKIA SMARTPHONE HACKS™

Michael Juntao Yuan

O'REILLY®

Beijing · Cambridge · Farnham · Köln · Sebastopol · Tokyo

Nokia Smartphone Hacks™
by Michael Juntao Yuan

Published by O'Reilly Media, Inc., 1005 Gravenstein Highway North,
Sebastopol, CA 95472.

O'Reilly books may be purchased for educational, business, or sales promotional use. Online editions are also available for most titles (*safaribooksonline.com*). For more information, contact our corporate/institutional sales department: (800) 998-9938 or *corporate@oreilly.com*.

Editor:	Brian Jepson	**Production Editor:**	Mary Brady
Series Editor:	Rael Dornfest	**Cover Designer:**	Hanna Dyer
Executive Editor:	Dale Dougherty	**Interior Designer:**	David Futato

Printing History:

July 2005:	First Edition.

ISBN: 0-596-00961-5
[LSI]   [9/14]

Contents

Foreword

You have in your hands *Nokia Smartphone Hacks*, from the gurus who deliver deep insight into anything cool. I am delighted to have been asked to write the foreword for this important book, which I believe will highlight details of some of the most open, extensible, and personalizable devices ever created.

Most people still think of phones as voice devices, which is natural, since they are so omnipresent. For me the phone ceased to be a phone when I started evangelizing in 1997 for a smartphone that could be operated with one free hand. In July of 2002, it came together with the launch of the Nokia 7650. I remember the November 23rd, 2002 cover for *The Economist* where the Nokia 7650 was set as the Monolith in Stanley Kubrick's *2001: A Space Odyssey*, in which primates learned to use clubs to shape their world. This picture was perfect and the headline said it all: "Computing's new shape." Figure P-1 shows artist Donald Smith's rendering of a similar image using the latest Nokia N91 smartphone.

> The Nokia N91 smartphone featured in the figure combines an MP3 music player, a mpeg4 video player, a 4GB hard drive (3,000 songs or many hours of high quality video), as well as a 2-megapixel digital camera and camcorder. It also features a large color screen, sliding keypad, and built-in 3G/WiFi/Bluetooth/USB data connectivity. It represents the future direction of smartphones.

Now, three years later, we are starting to see commercial success with the Nokia Series 60 platform built on top of the Symbian OS. There are more than 10 million Nokia Series 60 devices in the market, and the number is growing very rapidly. The platform is now being used for imaging, gaming, music, and video. I guess it would be fair to say that we've only just gotten started.

Figure P-1. The Dawn of Smartphone, by Donald L. Smith

An open platform such as the Nokia Series 60 brings the power to the individual: with a few simple actions, you can personalize the device exactly how you want it. In this book, you will find lots of great tricks and hacks that make a smartphone the first *intimate computer*.

An open platform brings power to developers; in this book you will find lots of insight into mobile devices essential for anyone entering the mobile application or service business.

Through faith I ended up being a developer on the Series 60 platform with the development of Nokia Lifeblog (*http://www.nokia.com/lifeblog/*). I am frequently amazed at what you can do when you have a fully programmable platform. I wish I would have had this book as a reference, since we had to dig out the details ourselves.

I am really pleased that Michael, Brian, and the team have undertaken this task of digging out the details of the world of Nokia smartphones. This should be standard reading for anyone shaping the future of computing.

—Christian Lindholm
"Father of the Nokia Series 60 user interface"

Credits

About the Author

Michael Juntao Yuan (*http://www.MichaelYuan.com/*) is a mobile alpha geek and author of three mobile-technology-related books. Over the last couple of years, he has managed to accumulate more than a dozen smartphones—most of them are Nokia phones. But he is still waiting for the ultimate smartphone that converges the functionality of a mobile phone, iPod, PDA, PlayStation Portable, prosumer digital camera, GPS, XM radio, TV, key card, credit card, and Swiss army knife. Of course, that device must also fit into his back pocket and have more than 24 hours of battery life in continuous operation. OK, maybe he is dreaming. But isn't mobile technology all about turning dreams into reality? In fact, with some hacking, a lot of this functionality can already be achieved on today's smartphones. You just need to read this book. :-)

Professionally, Michael specializes in end-to-end software solutions for enterprises. He is the lead developer of Nokia's Series 40 Java Blueprint Application and is actively involved in various standards committees helping to define the next-generation Java platform on mobile phones. Michael currently works for JBoss Inc., the Professional Open Source software company, focusing on next-generation Java enterprise middleware. He writes articles for leading journals and speaks at many conferences on subjects ranging from mobile application development to software development process management.

Michael has a Ph.D. from the University of Texas at Austin. When he is not working, writing, or playing with gadgets, he likes to travel and take pictures. Check out his picture albums at *http://www.jjcafe.net/photography/*.

Contributors

This book covers a diverse range of hacks, written by a diverse group of contributors. The following people contributed their hacks and knowledge to this book:

- Brian Jepson is an O'Reilly editor, programmer, and coauthor of *Mac OS X Tiger for Unix Geeks* and *Linux Unwired*. He's also a volunteer systems administrator and all-around geek for AS220 (*http://www.as220.org*), a nonprofit arts center in Providence, Rhode Island. AS220 gives Rhode Island artists uncensored and unjuried forums for their work. These forums include galleries, performance space, and publications. Brian sees to it that technology, especially free software, supports that mission.

- R. Emory Lundberg lives in Providence, Rhode Island, with his wife, Elizabeth, and a chubby calico cat named Echo(1). By day he applies ninja tactics to errant packets for VeriSign's excellent Managed Security Services team, and by night he writes about mobile technology and tinkers with gadgets. He will one day be survived by his motor scooters and an exquisite collection of mobile phones.

- Schuyler Erle was born near 39.9 N 75.2 W, and later on earned a degree studying linguistics nearby at Temple University. Sometime afterward, he moved west in search of dot-com fame and fortune, but instead ended up near 38.4 N 123 W, writing software for several different departments within O'Reilly Media. During his stint at O'Reilly, Schuyler got into wireless networking and, in his spare time, cowrote the NoCat Authentication System, one of the earliest and still one of the most widely used open source captive portal packages. Schuyler's interest in automating analysis of possible long-distance 802.11 links led to collaboration with Rich Gibson on the NoCat Maps project. Together with Schuyler's wife and collaborator, Jo Walsh, the three of them wrote *Mapping Hacks*, which was published by O'Reilly in 2005. Today, Schuyler and Jo spend most of their time around 51.1 N 0.1 W, but tomorrow they might well be found somewhere else entirely. Schuyler is, among other things, chief engineer of Locative Technologies, a consultancy offering expertise in open source GIS, with an emphasis on populist design, grassroots political and humanitarian action, and sustainable economic development.

- Rael Dornfest is chief technology officer at O'Reilly Media. He assesses, experiments, programs, fiddles, fidgets, and writes for the O'Reilly Network and various O'Reilly publications. Rael is series editor of the O'Reilly Hacks series and has edited, contributed to, and coauthored various O'Reilly books, including *Mac OS X Panther Hacks*, *Mac OS X Hacks*, *Google Hacks*, *Essential Blogging*, and *Peer to Peer: Harnessing the Power of Disruptive Technologies*. He is also program chair for the O'Reilly Emerging Technology Conference. In his copious free time, Rael develops bits and bobs of freeware, particularly the Blosxom weblog application (*http://www.blosxom.com/*), is editor in chief of MobileWhack (*http://www.mobilewhack.com/*), and (more often than not) maintains his Raelity Bytes weblog (*http://www.raelity.org/*).

- Edd Dumbill is a free software enthusiast, internet entrepreneur, and Englishman. He has a weblog at *http://usefulinc.com/edd/blog*.

- Haihao Wu grew up in Shanghai and lives in Austin, Texas with his wife. He works for Freescale Semiconductor.

- Ju Long is an assistant professor for Computer Information Systems at the McCoy School of Business at Texas State University. She has a Ph.D. degree from the University of Texas at Austin and a master's degree from the University of Michigan at Ann Arbor. She studies technology's impact on business and how new software development methodologies are changing our world.

- Trent Fitzgibbon works as a senior software engineer for Nokia Finland. He graduated from Monash University and was named "Outstanding Student in the Bachelor of Computer Science" by the Australian Computer Society in 1999. Since moving to Finland three years ago, he has been developing advanced smartphone software for Nokia's Series 60 platform. Previously, Trent worked in Australia and the U.S. as an embedded software engineer.

- Kamil Kapadia was born and raised in Concord, California. He graduated from the University of California, San Diego, with a bachelor of science degree in Computer Science. At a young age, he learned to program in BASIC and fell in love with programming. After Kamil was introduced to C++, he decided he wanted to become a programmer. Kamil currently resides in Northern California, where he spends his free time learning about new programming languages and technologies.

- Frank Koehntopp is an IT security professional and mobile geek. He's 38, married with two kids, and lives in Germany. His alter ego, The GadgetGuy, writes a weblog on *http://www.gadgetguy.de/*. He can be reached through the *#mobitopia* IRC channel on *irc.freenode.org*, where he often hangs out with other mobile geeks.

- Ajay Kapur is Founder and President of Moov Software, a developer of smartphone applications. Previously, Ajay worked at Goldman Sachs, making venture capital investments in semiconductor, systems, and software startups. Ajay has a BA in Physics and Computer Science from UC Berkeley, where he was a Regents' Scholar. He will be graduating from Stanford's Graduate School of Business in 2007.

- Donald Smith has been involved in digital image making for more than 10 years. He worked on the animated television series "Voltron the Third Dimension" for Netter Digital Entertainment and "Max Steel" for Sony Television and Warner Brothers. Donald has a BFA degree from the University of Texas at Austin, and is involved in painting, 3D animation, and video. He can be reached by email at *don_3d@yahoo.com* and has a online portfolio at *http://www.geocities.com/don_3d*.

Acknowledgments

First, I would like to thank Brian Jepson, my editor at O'Reilly Media. Brian helped me organize the book from the very beginning and provided key guidance throughout the development process. His thoughtful edits and sharp comments have greatly improved the quality and coverage of the book. Brian has also personally contributed several important hacks. It is a great privilege to work with you, Brian!

R. Emory Lundberg is the technical reviewer of the book. But his contribution is not limited to valuable review comments. Emory directly contributed several hacks. His writings in MobileWhack (*http://www.mobilewhack.com/*) and other places have inspired other hacks in the book. Thank you, Emory!

Haihao Wu is a personal friend of mine and a fellow mobile geek. He not only contributed two hacks, but also spent hours on the phone and in face-to-face conversations helping me solve problems. But most importantly, he permitted me to play with his N-Gage and PlayStation Portable when I was not writing. Haihao, DON'T PANIC.

I am indebted to all of those who contributed to this book, as well as to the following individuals who offered help and encouragement at various points during the writing process: Michal Bacik, Jonas Salling, Colleen Romero, Khoa Duong, Goldy Lukka, and Kok Seong Khew. I would also like to thank Forum Nokia for contributing a very useful hack on the Nokia Theme Studio.

Finally, I would not have archived anything without the loving support of my beautiful wife and collaborator, Ju Long. She has to endure my long working hours and busy weekends. As a noted mobile commerce researcher, Ju contributed two hacks directly to the book, and her professional opinions influenced many others. You are the best, Ju!

Preface

Smartphones are not only our smallest computers, but also our most connected computers. They connect to the voice telephone network as well as to the short- and long-range digital data network, including the Internet itself. In terms of computing power, a typical smartphone in 2005 is roughly equivalent to a top-of-the-line PC of the middle 1990s in terms of both CPU speed and memory size. By combining the most disruptive technologies of our times, including wireless networks, mobile phones, digital media, PCs, and the Internet, smartphones are profoundly changing our lives and our social interactions. It can be argued that there is a "mobility divide" similar to the "digital divide" in our societies: the people who know how to take advantage of smartphones will eventually have a competitive edge. If you are interested in learning how to make the best use of your smartphone, this book is for you.

In this book, a smartphone is defined as a mobile phone that has computer functionality, such as a web browser, an email client, a personal information manager, a media player, and video games. Under this definition, almost all Nokia mobile phones sold since 2003 are smartphones. If your Nokia phone has a color screen, it is most likely a smartphone covered in this book. Please note that our definition of smartphone is much broader than Nokia's own definition, which calls only its Symbian-based midrange to high-end phones smartphones.

Why Nokia Smartphone Hacks?

The term *hacking* has a bad reputation in the press. They use it to refer to someone who breaks into systems or wreaks havoc with computers as their weapon. Among people who write code, though, the term *hack* refers to a "quick-and-dirty" solution to a problem, or a clever way to get something done. And the term *hacker* is taken very much as a compliment, referring to someone as being *creative*, having the technical chops to get things done. The Hacks series is an attempt to reclaim the word, document the good ways people are hacking, and pass the hacker ethic of creative participation on to the uninitiated. Seeing how others approach systems and problems is often the quickest way to learn about a new technology.

Hacks are especially useful for Nokia smartphone users. A Nokia smartphone is not only a voice communications device, but also a fully featured computer capable of running third-party software. There are many creative ways (i.e., hacks) in which you can customize a smartphone, add features to it, and make it work best for you. Yet, few people know about these tricks and tips, as they often require you to dive deep into the user interface or download additional software. This book aims to bring those powerful hacks to all Nokia smartphone users. Read on and start the adventure!

How to Use This Book

You can read this book from cover to cover if you like, but each hack stands on its own, so feel free to browse and jump to the different sections that interest you most. If there's a prerequisite you need to know about, a cross-reference will guide you to the right hack.

If you are not very familiar with the Nokia product lines or are unsure of what phone you have, I recommend that you read "Choose the Right Phone for the Network" **[Hack #1]** and "Pick the Right Class of Nokia Phone" **[Hack #2]** first. They explain terms such as *Series 40*, *Series 60*, and *GPRS*, which are used frequently in the rest of the book.

How This Book Is Organized

The book is divided into several chapters, organized by subject:

Chapter 1, *Get to Know Your Phone*
Use the hacks in this chapter to understand the basic characteristics of your smartphone and its related mobile network services. If you do not have a Nokia smartphone already, some of the hacks can serve as a nice buyer's guide to help you choose from hundreds of combinations of devices and service plans. You will also learn advanced hacks such as how to unlock your existing smartphone when you switch operators and how to change network settings via operation codes.

Chapter 2, *Get Connected*
Smartphones are revolutionary devices because they can connect to the digital network anytime, from anywhere. In this chapter, you will learn how to connect your smartphone to the Internet and to nearby computers.

Chapter 3, *Extend and Enhance Your Phone*

Like a regular computer, smartphones can run third-party software programs. And as it has been proven in the computer world, software is often more important than hardware. You can download, purchase, or write your own programs to extend and enhance the functionalities of your smartphone. Many hacks in this book require you to install additional software on your device. Read this chapter carefully to learn how to use, manage, and even develop smartphone software.

Chapter 4, *Protect Your Phone*

A weakness of connected computers is the rise of viruses and malicious programs that spread over the network. Mobile phone viruses are already in the wild. In this chapter, you will learn how to prevent malicious programs from infecting your phone, and how to recover from them if they do. You will also learn how to deal with lost and stolen devices.

Chapter 5, *Make and Receive Voice Calls*

Making voice calls is and will continue to be the most important functionality of smartphones. A lot of smartphone software, including applications bundled in the smartphone operating system and third-party software, aim to make the smartphone a better telephone. This chapter covers interesting hacks such as speed dialing, recording phone conversations to digital files, and using calling cards.

Chapter 6, *Exchange Data with Computers*

The smartphone is your smallest computer, but it is not your only computer. In fact, the smartphone is most useful when it is paired together with a desktop computer that has a full-size keyboard, a large monitor, and more computational resources. The PC can act as the smartphone's content repository, and handle most of the computationally intensive data processing. To use the smartphone with the PC, the first thing you need to do is to move data from one device to another. In this chapter, you will learn all about data exchange and synchronization between smartphones and PCs.

Chapter 7, *Enhance the PC Experience*

While it is obvious that a PC can enhance a smartphone by providing offline computing power, a smartphone can also make the PC more useful. In this chapter, you will learn how to use a smartphone as a data modem to provide Internet access to computers anytime, anywhere. You will also learn how to control your computers remotely via your smartphone.

Chapter 8, *Improve the User Interface*

The smartphone is a highly personal device. You should carefully customize its user interface, including sound and graphics, to make it work best for you. In this chapter, you will learn innovative hacks to use the profile, ring tones, and graphics and fonts on the phone display.

Chapter 9, *The Mobile Web*

A key feature of smartphones is their ability to browse web sites on the Internet. They can access not only special mobile sites (i.e., WAP sites), but also general HTML sites as well as RSS-based blog sites. In this chapter, you will learn how to develop your own mobile web site, use mobile portals and search engines, and post to mobile photo blogs from your smartphone.

Chapter 10, *Email and Messaging*

With smartphones, you can send and receive email and mobile messages on the go. In this chapter, you will learn various ways to set up email and instant messaging on your smartphone. You will also learn how to use mobile email and SMS-based services efficiently.

Chapter 11, *Mobile Multimedia*

Most smartphones have cameras. Camera phones are among the best-selling mobile phones in the world. In this hack, you will learn how to take good photos and video clips with your phone camera and share them with friends across the world. The chapter will also cover how to play MP3 music and DVD movies on a Nokia smartphone.

Want to Learn More?

The technology landscape for smartphones is highly competitive and is evolving quickly. Nokia releases around 30 new models of mobile phones every year. Wireless operators are constantly competing with each other by providing new service offerings. The hacks in this book are generic enough to cover a wide range of future smartphones and mobile services. But occasionally, you might still need to read beyond the book to find information about a new smartphone or to troubleshoot a hack. I recommend the following popular web sites for further reading or research:

- Forum Nokia (*http://www.forum.nokia.com/*) is Nokia's official web site for mobile application developers. You can find a lot of technical articles, the latest device matrix, and software tools here. Forum Nokia also runs discussion forums where you can ask questions related to your Nokia phone.

- All About Symbian (*http://www.allaboutsymbian.com/*) is a web site for the Symbian smartphone community, including the Nokia Series 60 smartphone community. This site has the latest device news, and reviews of upcoming devices.

- Howard Forums (*http://www.howardforums.com/*) is a well-established online BBS for discussing mobile-phone-related issues. Its Nokia discussion boards are very popular.

- My-Symbian.com (*http://my-symbian.com/*) is a news and community discussion site for Symbian (Nokia Series 60) smartphones. It also features Symbian software and mobile content download.

- NokiaFree Forums (*http://nokiafree.org/forums/*) features forums on Nokia hardware repairs and flash memory reverse engineering.

- Mobitopia (*http://www.mobitopia.com/*) is a community blog for mobile technology. It is operated by a group of mobile phone enthusiasts and power users. Mobitopia is frequently updated (several times a day) and mainly features news analysis by its members.

- Mobile Burn (*http://www.mobileburn.com/*) is a news and discussion site for mobile phones. It has some of the best and most comprehensive reviews for upcoming Nokia smartphones. Plus, it has lots of pictures!

- SymbianOne (*http://www.symbianone.com/*) has articles, reviews, discussion forums, and content downloads for Symbian smartphones (Nokia Series 60).

- Handango (*http://www.handango.com/*) is an online store for mobile software. It has a large selection of Java and Symbian applications, as well as mobile content. You can select compatible software and content based on your phone model.

- Engadget (*http://www.engadget.com/*) is a managed blog site reporting the latest news in electronic gadgets. It is very frequently updated (sometimes 50 news items per day), and it has extensive coverage of Nokia smartphones.

Links to the preceding resources and other smartphone-related sites and blogs can be found at the book's web site at *http://www.MichaelYuan.com/NokiaHacks/*.

Conventions

The following is a list of the typographical conventions used in this book:

Italics

Used to indicate URLs, filenames, filename extensions, and directory/folder names. For example, a path in the filesystem appears as */Developer/Applications*.

Constant width

Used to show code examples, the contents of files, console output, as well as the names of variables, commands, and other code excerpts.

Constant width bold

Used to highlight portions of code, typically new additions to old code.

Constant width italics

Used in code examples and tables to show sample text to be replaced with your own values.

Color

The second color is used to indicate a cross-reference within the text.

You should pay special attention to notes set apart from the text with the following icons:

This is a tip, suggestion, or general note. It contains useful supplementary information about the topic at hand.

This is a warning or note of caution, often indicating that your money or your privacy might be at risk.

The thermometer icons, found next to each hack, indicate the relative complexity of the hack:

 beginner   moderate   expert

Using Code Examples

This book is here to help you get your job done. In general, you can use the code in this book in your programs and documentation. You do not need to contact us for permission unless you're reproducing a significant portion of the code. For example, writing a program that uses several chunks of code from this book does not require permission. Selling or distributing a CD-ROM of examples from O'Reilly books *does* require permission. Answering a question by citing this book and quoting example code does not require permission. Incorporating a significant amount of example code from this book into your product's documentation *does* require permission.

We appreciate, but do not require, attribution. An attribution usually includes the title, author, publisher, and ISBN. For example: *"Nokia Smartphone Hacks* by Michael Yuan. Copyright 2005 O'Reilly Media, Inc., 0-596-00961-5."

If you feel your use of code examples falls outside fair use or the permission given here, feel free to contact us at *permissions@oreilly.com*.

How to Contact Us

We have tested and verified the information in this book to the best of our ability, but you might find that features have changed (or even that we have made mistakes!). As a reader of this book, you can help us to improve future editions by sending us your feedback. Please let us know about any errors, inaccuracies, bugs, misleading or confusing statements, and typos that you find anywhere in this book.

Please also let us know what we can do to make this book more useful to you. We take your comments seriously and will try to incorporate reasonable suggestions into future editions. You can write to us at:

O'Reilly Media, Inc.
1005 Gravenstein Highway North
Sebastopol, CA 95472
(800) 998-9938 (in the U.S. or Canada)
(707) 829-0515 (international/local)
(707) 829-0104 (fax)

To ask technical questions or to comment on the book, send email to:

bookquestions@oreilly.com

The web site for *Nokia Smartphone Hacks* lists examples, errata, and plans for future editions. You can find this page at:

http://www.oreilly.com/catalog/nokiasmarthks

For more information about this book and others, see the O'Reilly web site:

http://www.oreilly.com

Got a Hack?

To explore Hacks books online or to contribute a hack for future titles, visit:

http://hacks.oreilly.com

Get to Know Your Phone
Hacks 1–9

Most casual mobile phone users use their phones only for voice calls. The phone's keypad and small screen are well designed for such use. However, the modern Nokia mobile phone is much more than a voice communications device. In fact, the phone comes with a 200-page user manual, which covers many of the phone's advanced features.

Today's Nokia mobile phone is a voice, text, and multimedia communications hub, a personal information manager, a web browser, a multimedia entertainment center, a game console, and an extensible platform for new applications. In the rest of this book, I will discuss tricks and tips on how to make the best use of those features. But before you start on those hacks, you need to understand the Nokia phone's features and how they relate to network services. That's the focus of this chapter.

> As I discussed in the Preface, in this book a smartphone is defined as a mobile phone that has computer functionalities, such as a web browser, an email client, a personal information manager, a media player, and video games. Under this definition, almost all Nokia mobile phones sold since 2003 are smartphones. If your Nokia phone has a color screen, it is most likely a smartphone covered in this book. Please note that our definition of smartphone is much broader than Nokia's own definition, which calls only its Symbian-based mid-range to high-end phones smartphones.

The best place to find information about Nokia phones is the Nokia web site. Unfortunately, the Nokia web site is hard to navigate and bookmark, since most links are generated dynamically. If you are curious about a specific model, you can just enter the word *Nokia*, along with the four-digit number of the model you're interested in (e.g., *Nokia 6600*), into the Google search engine (*http://www.google.com/*). The Nokia product page is typically the top-ranked search result. From that page, you can learn more about the phone, read the technical specifications, download the user manual, obtain additional software, and buy accessories.

 H A C K
 # 1

Choose the Right Phone for the Network

A Nokia mobile phone works with either the GSM network or the CDMA network. You need to choose the right model to match your wireless operator.

Not all Nokia smartphones work with all wireless operators everywhere in the world. Wireless operators build their networks using different technologies, and a given phone typically works with only one of the network technologies.

If you got your Nokia smartphone directly from the wireless operator when you signed up for a service plan, you can rest assured that the device is compatible with the operator's wireless network in your area. However, as smartphone power users and gadget lovers, many of us actually buy cutting-edge devices directly from electronics stores and then use them with existing service plans. In this case, you need to be a little careful to select a phone that works with your operator's network.

 Buying a smartphone from a third-party electronics store without a service plan will usually cost you more, since the mobile operator does not subsidize the cost of the device with service revenues in this case. However, you will have access to the latest devices that have not been officially supported by the operator, and you will be freed from operator lock-in [Hack #7]. Additionally, you get a handset unmolested by the mobile operator. Many operators apply custom firmware and branding to their handsets, such as adding new icons or removing functionality from the device. Even if you get a device unlocked from your operator, it will still have these customizations in place, and sometimes this can produce very undesirable effects.

Popular networking technologies used by today's wireless operators include GSM (Global System for Mobile communications), iDEN (Integrated Digital Enhanced Network), and CDMA (Code Division Multiple Access):

GSM

The GSM technology was originally developed in Europe for digital voice networks. It has been extended to support fast wireless data access via the General Packet Radio Service (GPRS; 20–40 kbps), EDGE (40–230 kbps), and UMTS (384 kbps and up) protocols. GSM phones operate at frequencies of around 900, 850, 1800, and 1900 MHz. It is now the most widely used wireless phone network technology in the world. Most Nokia phones support GSM and GPRS. A few newer models also support EDGE and UMTS. I discuss how to set up data networking on Nokia GSM phones in "Connect Your Phone to the Internet" [Hack #10].

iDEN

The iDEN technology is Motorola's proprietary wireless networking technology based on (but not compatible with) GSM. It attempts to combine the digital phone, two-way radio, alphanumeric pager, and data/fax modem in a single network. The iDEN network is used in about a dozen countries. In North America, Nextel and Mike (Telus Mobility) are iDEN operators. Nokia phones do not work on iDEN networks.

CDMA

The CDMA technology is developed and licensed by Qualcomm. It is a spread-spectrum technology that uses a single-frequency band for all traffic. Individual transmission is differentiated via a unique code assigned by the network before transmission. CDMA supports wireless data at a peak of 153 kbps, roughly three times the speed of GPRS. As of late 2004, 15 models of CDMA-compatible Nokia devices were available. However, when it comes to device software and additional features, Nokia's CDMA devices are not as technically advanced as its GSM-based devices.

Network availability is a consideration if you frequently travel abroad. For instance, most wireless operators in Western Europe use the GSM network. Your CDMA phone will have limited coverage there. On the other hand, South Korea uses mostly CDMA networks.

To get a complete list of GSM and CDMA devices from Nokia, go to the Forum Nokia device matrix at *http://www.forum.nokia.com/devices* and choose GSM or CDMA in the "Devices filtered by" drop-down box (see Figure 1-1).

The TDMA filter in the device matrix lists older TDMA devices. Those devices are no longer actively marketed and are not compatible with the newer GSM networks.

Pick the Right Class of Nokia Phone

Nokia phones come in many shapes, sizes, and models. The Nokia Device Series provides a logical way to categorize those phones according to their features.

In 2004, Nokia announced more than 40 new models of mobile phones to satisfy the diverse needs of the mass market. Add in the older devices that are still being supported, and you've got about 100 different models of Nokia phones currently in use by consumers. For most users, this forest of

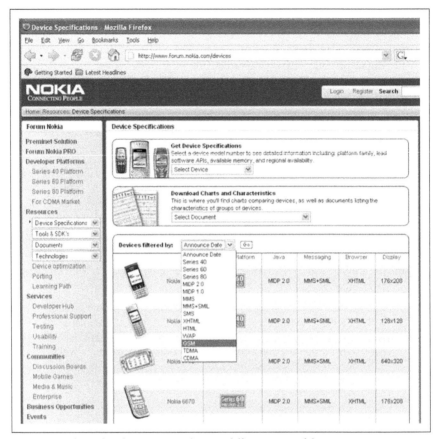

Figure 1-1. The Nokia device matrix showing different types of devices

devices is confusing and difficult to navigate. Here is where a little knowledge about the Nokia device series could really help. If you are considering buying a Nokia smartphone in the near future, this hack also serves as a buyer's guide.

Most Nokia devices share many common characteristics and can be grouped into several device series. The Nokia device series distinction is primarily a convenience for application developers (who need to distinguish between classes of device capabilities) rather than general consumers. Yet, as power users, knowledge about the device series helps us quickly identify the user interface (UI) style, available software, and hacking options on any Nokia device. In this book, I discuss hacks in the context of device series. For example, some hacks are applicable to only one series of devices, and others need to be applied differently on each series.

Currently, most Nokia devices are grouped into three series: Series 40, Series 60, and Series 80. You can see a complete list of devices in each series by filtering through the device series in the Forum Nokia device matrix [Hack #1]. Now, let's look at the key characteristics of the devices in each series.

> The device series information is especially useful when you need to purchase a new device. Based on your software and UI requirements, you can quickly narrow down your search to a specific series. Then you can dig into the detailed device specification via the Forum Nokia device matrix or the consumer home page for each device on the Nokia web site.

Series 40

Series 40 devices are mass-market consumer devices with hundreds of millions of users. Figure 1-2 shows several popular Series 40 devices in use today.

A typical Nokia Series 40 device features a 128×128 LCD display with a 12-bit color depth (4,096 colors). Some devices have 96×65 or 128×160 LCD screens and other color depths. The UI on Series 40 devices consists of a hierarchy of view-switch screens. It typically displays five lines of text, plus headers. The keypad has the traditional alphanumeric keys, a four-way scroll key, the Send/End keys, and two or three generic soft keys. The device displays images in common file formats, receives AM/FM radio station signals, records voice messages, and plays Musical Instrument Digital Interface (MIDI) polysynthetic ring tones. Device extensions such as cameras, full alphabetic keyboards, and MP3 players are available on selected Series 40 device models.

> Nokia does not officially call Series 40 devices "smartphones." However, as I indicated, Series 40 devices have a lot of smartphone features. Series 40 devices are covered in this book.

The software on Series 40 devices is based on the proprietary Nokia OS. The device is shipped with native software for telephony functions (e.g., the call log), Personal Information Management (PIM—e.g., contact lists and calendar), messaging (e.g., SMS, MMS, and email), and web browsing (e.g., WAP), etc. Since the Nokia OS is not open to developers outside of Nokia, you cannot hack into the core of the phone's operating system. However, Series 40 devices do have limited programmability via their Java application environment. So, you can hack and enhance the phone with Java applications [Hack #16].

Figure 1-2. Popular Nokia Series 40 devices

Many highly successful Nokia Series 40 devices are available. The Nokia 7210 is one of the first compact color phones to support MMS and WAP browsing. The Nokia 6230 features a camera and advanced device software. The Nokia 5140 is a rugged device designed for an active lifestyle. The Nokia 6820 features a full keyboard and a camera. The Nokia 6170 is one of Nokia's first flip-open models.

Series 60

The Nokia Series 60 devices target the midrange to high end of the consumer market. Figure 1-3 shows several popular Series 60 devices.

Figure 1-3. Popular Nokia Series 60 devices

A Nokia Series 60 device typically has a 176×208 LCD screen capable of displaying 65,536 (16-bit) colors. Series 60 devices with other UI configurations (e.g., the 640×320 touch screen on the Nokia 7710) were starting to emerge as of early 2005. The UI of a Series 60 device is similar to that of a PDA, with icons in the grid layout and standard menus from the toolbar. Compared to a standard Series 40 keypad, a Series 60 keypad has several additional keys, including an Application key, a Clear key, and an Edit key. A Series 60 device plays Audio/Modem Riser (AMR) voice tones as well as other Series 40 audio formats.

The Series 60 platform is a licensable product from Nokia. It is licensed to seven other device makers.

Similar to Series 40 devices, Series 60 devices ship with native applications for making phone calls, PIM, messaging, web browsing, and more. However, a major difference between Series 40 and Series 60 devices is in the base software. Instead of the proprietary Nokia OS, Series 60 devices are based on the Symbian OS, which exposes core device functionality via the open Symbian C++ API. "Run Symbian Applications" [Hack #17] covers how to install Symbian native applications onto your Series 60 device. In addition, Series 60 devices are programmable via the Java environment. As a result, the Series 60 devices are much more "hackable" than the Series 40 devices. Some of the hacks in this book apply only to Series 60 devices.

It is important to note that two different versions of the Symbian OS are in use today on Nokia Series 60 devices. Older devices, such as the Nokia 3650, use Symbian OS 6. Devices released after 2003, such as the Nokia 6600, use Symbian OS 7. The UIs of the two OS versions are slightly different. So, throughout the book, I sometimes differentiate between "newer" and "older" Series 60 devices.

Nokia Series 60 devices are among the best-selling smartphones in the world. The Nokia 3650 camera phone is very popular. The newer Nokia 6600/6620 camera phones feature hardware and software improvements over the Nokia 3650. The Nokia N-Gage is the first smartphone that is also a mobile game console.

Series 80

The Nokia Series 80 devices target high-end enterprise users. A Series 80 device typically has two user interfaces. When the phone is unfolded, a large 640×200 screen and a full alphabetic keyboard become available. You can use the phone as a sort of mini-laptop computer with this setup. When the phone is closed, the external 128×128 color screen and keypad provide a Series 40-style user interface.

The Series 80 devices are based on the Symbian OS. However, since its UI style is different from that of the Series 60, there is no guarantee that Series 60 Symbian applications will run on Series 80 devices. The hacks in this book are not tested on Series 80 devices.

Series 80 devices include the Nokia 9500 and 9300 Communicators.

Series 90

If Nokia had a secret lab in a mountain somewhere, with giant lasers and an evil genius, it probably used this lab to make its Series 90 devices. Nokia originally planned to release a device called the 7700, which featured a wide-angle display and a touch screen. For whatever reason, it sacked this handset and then released the 7710, which is a very unique multimedia powerhouse. The larger display is great for video and for Nokia's new "Visual Radio" that it's testing in Finland, but the Series 90 platform has been swallowed up by other development efforts, and much of what made the Series 90 so interesting will be incorporated into the Series 60 in the future. If you have a Nokia 7710, you have an island of Nokia technology, albeit a very cool island.

Pick a Voice Plan

Given a choice between CDMA and GSM wireless network operators, how do you decide which kind of network, let alone which operator, to trust with your calls and mobile data?

We all love the gadgetry and cool factor of new equipment coming out these days. However, *gizmophilia* shouldn't be your primary deciding factor when picking a wireless network operator (a.k.a. carrier). Unless you live in a major market comprising at least a couple million people and choices and coverage galore, there are more important considerations than the quality of the onboard PIM or the choice of downloadable games. A mobile phone is simply no good without service, and it's fairly useless and more than a little irritating with spotty or iffy service. You should focus first and foremost on the available coverage and spectrum in your most frequent markets of use. Use the operator's web site to look up your home and your office, and if you regularly visit another city or town, check that out too. Make sure the coverage map of the network operator you are looking at covers the areas you frequent. Nothing is worse than discovering that you can't get a signal in the areas you visit regularly, or that you're subject to large roaming fees!

Just so that I'm perfectly clear: there is no such thing as a *best* wireless operator. It really isn't possible to be the best all around, for a lot of reasons, but I hope to give you the means to make an informed decision about which operator you should use in your primary market.

The two networks we are most concerned with are GSM and CDMA [Hack #1]. TDMA and AMPS are no longer being deployed, and Nokia does not make phones that support iDEN networks.

Pick the Phone

I recommend you find a handset [Hack #2] that answers your needs, and then find an operator that can support that handset or an operator that sells that handset at a discount. The CDMA operators typically aren't very willing to let you move to them a CDMA handset that you bought from a competing operator, but it has been known to happen (it can't hurt to ask). You'll have better luck moving from operator to operator with a GSM handset [Hack #7], because you can usually swap a small card called a *SIM card* from one phone to another; the SIM card identifies you and authorizes you to access a GSM network operator's network.

Once you've picked a phone, you'll be able to figure out from its description which kind of network it uses (CDMA or GSM). And once you've figured out which kind of network you need, you can use WirelessAdvisor to find out which operators in your area operate that kind of network. The search feature at *http://www.wirelessadvisor.com/* can look up which operators have spectrum in any major city or Zip Code in the U.S. This doesn't necessarily mean these companies are providing service; just that they own radio spectrum licenses there (most, if not all, of the listed operators provide service, although some of the operators who have licenses in a given area might not offer service yet). They will also list what kind of network they use, be it GSM, CDMA, iDEN, or TDMA.

Pick the Operator

Suppose you really want to use a Nokia 3660 smartphone. This is a GSM handset, so entering your Zip Code into the WirelessAdvisor search form gives you all operators that have GSM radio licenses in your area. Figure 1-4 shows the WirelessAdvisor with partial results for the 02881 (Kingston, which is located in Washington County, Rhode Island) Zip Code. Two operators are shown in that figure (Verizon and T-Mobile), although Verizon is broken out into its 800MHz and 1900MHz networks. So, if you see a provider on the list, such as T-Mobile, you can check the T-Mobile online store or a reseller to see whether that provider carries the phone (or a similar model).

At this point, you need to start listening to other customers' experiences with the operator. Ask your family, friends, and co-workers which operator they use. They'll tell you whether they hate the service, and you'll know from personal experience whether they drop calls all the time when speaking to you on their mobile. This can be a sign of bad equipment, an overused network, or weak coverage in your area. I used one of the CDMA operators for years in the Washington, D.C., market. After decent service for

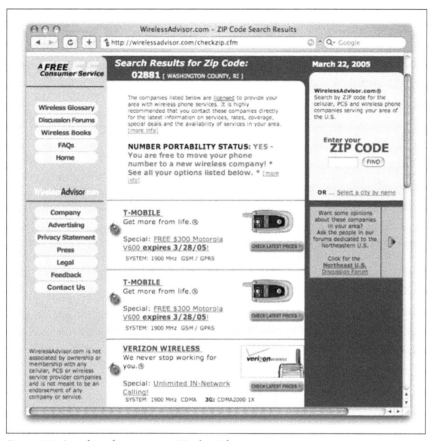

Figure 1-4. Searching for service on WirelessAdvisor

a while, they quickly went downhill. I regularly got "network busy" messages on my handset, I got dropped-call credits on my bill each month, and I had a very long and distressing call history with the customer service representatives. In spite of using this operator for years (nearly five, in fact), I could no longer think of them as reliable, so I ran like the wind to someone else. And I made it very clear to anyone I knew what I felt about that company. For every network operator, I hear plenty of complaints, and I've found that the quality of service depends not only on the operator, but also on how well that operator's network is implemented in a given geographic region.

So, you've tentatively selected an operator for your service with the phone you want (or one that you already own). Go to the operator's web site and check out its coverage map. Usually these maps are not hugely accurate, but if it's any consolation, operators generally err on the side of caution. See if the areas you frequent most are covered. If they aren't, check out another operator that can support the same equipment. Failing this, reconsider your equipment options or lower your standards. At least you can't say you didn't know your new phone wouldn't work in Hatteras, N.C. (a lovely place to visit, but last time I was there I had no mobile phone service of any kind).

Research Your Choices

To help you do some serious research on your market and the service that's available, WirelessAdvisor provides discussion forums. In addition, Howard Chui's excellent HowardForums (*http://www.howardforums.com/*) is where people talk about all things mobile. You can use the search feature at HowardForums in a provider-specific forum, for example, to look for information on your market (simply navigate to that forum and use the search box at the bottom of the page). Make sure you use the search function; it's a very busy site, and I'm willing to bet the question you want to ask has been asked before.

HowardForums also has forums for manufacturers of all the mobile phones available today; in fact, Nokia has a following at HowardForums. It is also a good place to learn about new plans and specials before the operator makes a formal announcement. Plus, all the operators have several employees who read these forums regularly. I've seen many mobile users get assistance there when they thought they were getting the runaround on the phone with customer service.

Choose a Plan

Network operators change their plans constantly and offer promotions frequently. However, these promotions are usually based on two-year contracts, and the only thing you usually get a deal on is the price of the phone. If you factor in the early termination fee (ETF), which is usually $175 to $200, the deal isn't that great. You'll be hard-pressed to find anyone who is thrilled about signing up for a two-year contract: usually one year and one day into it, they begin wishing they'd gone with the one-year deal. As a result, many folks break their contracts, opting to pay the ETF so that they can switch to an operator who has a plan or data service **[Hack #4]** they prefer.

So, look for a promotion that includes permanent bonus minutes or greatly discounted monthly rates. These plans do exist. When AT&T Wireless launched its GSM service, its $39.95 500-minute Charter plan included an extra 500 minutes that a friend of mine still has active on a line to this day.

Many of these promotional plans disappear if you switch to a different plan. You will likely find that you can't switch back to the promotional plan.

As far as which plan to choose, consider a minimum of 1,000 minutes if you're a business user who is going to use the phone for a lot of calls during the day. If you're using the phone for personal use, you can get away with far fewer minutes, such as 300 to 600, if your operator throws in free nights and weekends (and assuming you can be flexible in terms of when you use your phone).

Every operator has a different definition of nights and weekends, and these definitions change occasionally. Over the years, operators have defined nights as starting at 9:00 p.m. and weekends as beginning on Friday at midnight, giving you only two days of weekend use. On the flip side, some have defined weekends as starting on Thursday at midnight, and so include all of Friday, Saturday, and Sunday. In that case, a business user might be able to get away with a plan with a low number of minutes, assuming he can schedule all his conference calls on Fridays.

Once you've chosen a plan, it doesn't stop there. Keep an eye on the number of minutes you're using [Hack #6] and switch plans if you need to. If you switch plans, make sure you know whether doing so will force you into another contract term.

After You Sign Up

Most network operators give you a trial period of 15 to 30 days. Use those days to your utmost advantage! Go to your home, your office, and any social place you frequent and you anticipate needing good service. Call customer service. Call them multiple times and speak to different representatives. In these cases, if the service doesn't work as you want it to, you can drop out of your contract and wash your hands of the whole ordeal.

You will also get all manner of horror stories about the customer service of various operators. Customer service needs to improve across nearly all the operators. Customer service representatives (CSRs) need to be better trained in general, and specially trained to manage complaints and service issues. I think the general feeling is that they have you for the length of your contract, and that if you've been a customer for a while you are tied to the telephone number they assigned to you, so they don't worry about losing you.

However, the relatively recent ability to port your number from one operator to another may cause some of the deadbeat customer service departments out there to rethink that assumption.

 I've found that praising a helpful CSR to her supervisor has benefited me more (in terms of free minutes and other bonus items) than complaining about CSRs that haven't been so helpful. No doubt a case of catching more flies with honey.

I know I can be a pain sometimes, but I will always notify a supervisor when a representative has been especially considerate, polite, and helpful. I feel it's more important to give a high-five to the CSRs who are really pulling their weight in the trenches than it is to complain about the ones who aren't.

—Emory Lundberg

HACK #4   Choose a Data Plan

With their comprehensive wireless support, Nokia phones are some of the most data-friendly gadgets around. If you're going to use your phone to connect to the Internet, make sure you've got the right data plan.

Your Nokia phone can connect you to the Internet in a lot of ways, but it's going to cost you. Before you commit to a costly data plan, you need to consider what you'll do with the phone. Will you send and receive email on the phone [Hack #60]? Do you plan to send a lot of camera phone pictures or video clips [Hack #69]? How about posting to your weblog [Hack #56]? And most important, will you use your Nokia phone to connect (*tether*) your laptop to the Internet [Hack #40]? It's possible for you to use your phone for a lot of what your computer can do, and yes, there are plenty of hotspots around for that data fix when you need it. But I think you'll give in and tether eventually—just wait until the first time you open up your laptop and find no WiFi signal while your phone is showing five bars!

 We will use the word *tether* even when referring to wireless methods of using your handset, such as the Bluetooth connection.

Data plans come in two flavors: metered and unlimited. With a metered plan, you get anywhere between 1 and 20 megabytes per month as a base allotment, and if you go over, you pay by the kilobyte. There are two types of unlimited plans: handset plans and really, honest, we-mean-it unlimited plans that let you use your phone as a wireless modem for your laptop or PDA.

Flavors of Cellular Data

The phone typically relies on the wireless operator's data network to connect to the Internet. The underlying wireless network technology is known as the *data bearer* of the mobile data. Here's a quick overview of the data bearers available from today's wireless service plans. Since the vast majority of Nokia devices are GSM-based, I focus the discussion on that technology.

GSM

> The GSM network provides data access over the phone (data calls). It works in the same way as the dial-up modem on a PC. The benefit of this technology is that it is available everywhere you can get coverage. However, the drawbacks are the slow data rate (between 9.6 and 13.2 kbps) and the dedicated phone call for the connection. Since the call must be connected for the entire data session, it counts against your airtime minutes. If you want to make a voice call, you must first disconnect the data call. All Nokia Series 40 and Series 60 devices support the GSM data bearer. GSM data is known as 2G (second generation).

GPRS

> The GPRS network allows the phone to have direct access to the packet-switched data from the network. The phone does not need to dial any calls. This feature allows the phone to have always-on access to data without using up airtime minutes. GPRS is known as 2.5G (halfway between second and third generation). The data speed of 2.5G data bearers (between 20 and 40 kbps) is also faster than that of the 2G data bearers. All Nokia Series 40 and Series 60 devices support the GPRS data bearer.

EDGE

> The EDGE network works much like the GPRS network, except that EDGE has a much faster data rate (up to 230 kbps). The EDGE network coverage is still limited. In the U.S., only AT&T Wireless (now part of Cingular Wireless) has national EDGE coverage. Only a handful of Nokia devices released after 2004 support the EDGE data bearer.

UMTS

> The UMTS network is what's known as third-generation (3G) wireless networking technology. It offers broadband data speeds of around 384 bps. However, UMTS coverage is very limited. In the U.S., it is currently available in only a few selected cities. In addition, UMTS service plans can be very expensive. New Nokia phones such as the Nokia 6630 support UMTS.

Which data bearer is available to you depends on your wireless operator, your location, and your service plan. GPRS service should be available from all GSM operators wherever you have voice coverage. If your current service plan does not include any data service, you can call up your wireless operator and add it for an extra monthly fee. The data service is typically metered by the bandwidth you use in a billing period. Most operators also offer flat-fee subscriptions for unlimited data use. For instance, in the U.S., T-Mobile charges between $20 and $30 per month for its GPRS-only network, and Cingular charges $80 per month for its EDGE and GPRS networks.

Similar data bearers also exist in CDMA-based networks and are available on Nokia CDMA phones (Series 40). CDMA's data bearer is Single Carrier Radio Transmission Technology (1xRTT) and Single Carrier Evolution Data-Only (1xEV-DO). 1xRTT has typical rates of 70–120 kbps, and 1xEV-DO has typical downstream speeds of 300–500 kbps (with upstream speeds the same as those of 1xRTT).

Figure Out What You'll Need

This could be the hard part, but the good news is that you can just take a best guess. If the plan you select doesn't work out, most providers will let you change data plans midstream. Here are some considerations:

Web browsing

If you're going to browse the Web using your Nokia's built-in browser, chances are good that you won't use a lot of data, simply because that browser doesn't offer the rich graphical and multimedia content you get with a desktop browser—in this case, it might be OK to go with a metered plan. However, if you use a third-party browser, it's possible you could end up using a lot of data, especially if you find yourself turning to your phone for web surfing **[Hack #50]** more than you do your laptop.

Email

If you plan to use your phone for sending and receiving email, this won't take up a huge amount of data (but see the next item). For the most part, email is text, and unless you're likely to use your phone to work with office documents **[Hack #39]**, you probably won't move a lot of data with email. As such, you could get away with a metered data plan. If you're going to be sending and receiving lots of photos and video, get yourself an unlimited plan.

Tethering

If you plan to connect your laptop to your cell phone to get online, welcome to what some folks think is a gray area. Here's the problem: most of the low-priced unlimited data plans are intended for use with your phone only. However, it's technically possible to connect your laptop to your phone and get online. You will probably get away with this if you don't use a lot of data. However, anecdotes abound concerning people who claim to have received nastygrams from their cellular operator after using large amounts of data in this way. It's insanely simple for a cell provider to distinguish between traffic that originates from a phone and traffic that originates from a laptop. For example, every web browser transmits a User-Agent identifier every time you load a page: this is a dead giveaway.

Some providers will bill you differently based on your usage. AT&T Wireless (which was being absorbed into Cingular at the time of this writing) had a $24.99 monthly handset data plan, but vowed it would charge $1 per megabyte to users who tethered a laptop or PDA to their mobile phone. Furthermore, cell providers routinely, and sometimes temporarily, block access to certain ports (there have been reports of SSH, secure IMAP, and POP for email, and even secure HTTP being blocked), with the (misguided) rationale that most handset users don't need those ports.

If you plan to use your cell phone as your laptop or PDA's lifeline to the Internet and you don't want to risk unexpected overages or a service disconnection, go with an unlimited plan that explicitly supports tethering.

SMS and MMS
In theory, Short Message Service (SMS) and MMS come out of a different billing bucket than does Internet data, so the number of messages you send and receive shouldn't affect your choice of a data plan.

All those guidelines aside, the best thing to do is choose an unlimited data plan, if one is available (otherwise, pick the most generous metered plan). That way, if you don't use a lot of data, you can progressively downgrade each month. If you do use a lot of data, at least you won't get whacked with per-kilobyte charges.

Compare the Plans

Although wireless data might seem as though it's brand new, its pricing is settled, for the most part. Usually you'll pay around $5 to $10 per month for something that lets you do basic web surfing and email, and around $20 to $30 for more capabilities. So, if you're planning to use a lot of third-party network applications (such as instant messaging [Hack #66] and RSS readers [Hack #55]), you should go for the plan that gives you more. If you plan to use your phone as a wireless modem for your laptop [Hack #40], you should definitely choose an unlimited plan that supports tethering. Table 1-1 shows a few unlimited data plans that were current as of this writing.

Table 1-1. Unlimited data plans by provider

Provider	Handset	With tethering
Cingular	$24.99; Media Net Unlimited	$79.99; Data Connect Unlimited
Nextel	$19.99; Enhanced Data Service Plan	$54.99; Unlimited Wireless PC Access Plan
Orange UK	£88.13; (if you exceed 1000 MB per month, you will probably get a nastygram)	Same as handset
Sprint	$15; Unlimited Vision	None
T-Mobile	$4.99; Unlimited T-Zones (email and web only) $9.99; Unlimited T-Zones Pro	$29.99; T-Mobile Internet ($19.99 with a qualifying voice plan)
Verizon Wireless	$15; VCAST $4.99; Mobile Web (uses up plan minutes)	$79.99; BroadbandAccess or NationalAccess Unlimited

If you think Table 1-1 looks very U.S.-centric, you're right. Although the U.S. lags behind in terms of the latest gizmos, it's a feeding frenzy for those who are determined to get all-you-can-eat data. In other parts of the world, metered data plans are more common. And the U.S. has its share of those as well. Table 1-2 shows some of these plans, from the low-end to the high-end offerings, and includes the range of charges you can expect if you go over the metered limit.

Table 1-2. Metered data plans by provider

Provider	Low-end metered	High-end metered	Overage
Cingular	512 KB for $2.99	60 MB for $59.99	Varies with plan: $0.005/KB to $0.01/KB
Nextel	1 MB for $9.99	100 MB for $99.99	Varies with plan: $0.003/KB to $0.01/KB
Orange France	5 MB for €5	20 MB for €20	Varies with plan: €0.10/KB to €0.15/KB
Orange UK	4 MB for £4	400 MB for £52.88	Varies with plan: £0.59/MB to £1/MB
Rogers (Canada)	256 KB for CAN$3	100 MB for CAN$100	Varies with plan: CAN$1.02 to CAN$10/MB
Sprint	Unlimited plans only	N/A	N/A
Telcel Mexico	1000 KB for MEX$100	50,000 KB for MEX$500	Varies with plan: MEX$0.10/KB to MEX$0.02/KB
T-Mobile	Unlimited plans only	N/A	N/A
T-Mobile Germany	5 MB for €5	500 MB for €110	Varies with plan: €0.80/MB to €3.90/MB

Table 1-2. Metered data plans by provider (continued)

Provider	Low-end metered	High-end metered	Overage
Verizon Wireless	20 MB for $39.99	60 MB for $59.99	Varies with plan: $0.002/KB to $0.004/KB
Vodafone UK	2.6 MB for £2.55	51.1 MB for £34.04	£2/MB

In most cases, you'll choose the data plan when you sign up for your voice service. However, most providers will let you add or change your data plan at any time. But before you make a change, ask the all-important questions: will this require me to agree to a contract extension, and will I be charged a termination or activation fee to make this kind of change? If you don't like the answer to either question, you should reconsider adding a data plan until it's time to renew your contract, or look into setting up a data plan on a separate line of service.

> One problem you might run into is a customer service rep that is unfamiliar with the plan you want. The best thing you can do is make sure you know the name of the service. To find this out, visit the provider's web site. If you can't find the name of their data plan in five minutes, go through this simple exercise: add a phone to your shopping cart and then choose a voice plan—preferably, the same one you are signed up for. The data plans available to you might depend on what level of voice service you have. For example, the price for T-Mobile's unlimited Internet plan is $19.99 with most voice plans, and $29.99 without a voice plan (or if you have a cheap plan).
>
> At this point, the provider's web ordering system should offer you a data plan (you might need to click around for an optional services or features link), and you should write down the name of the service, call customer service, and ask them to add it to your plan.

Once you've selected a data plan, log into your provider's web site every day or so and keep an eye on your data usage as you go—if you accidentally configure your email client to check mail every 30 seconds, you could be in for a surprise. The data total that shows up on the web site will probably be behind by 24 hours, perhaps more if you've used your data plan while roaming.

If you're on a metered plan and see yourself getting dangerously close to the limit, change plans right away. Although most providers will prorate the new plan after you change it, make sure you understand what exceptions are in place between the date you change the plan and the date your billing cycle resets. For example, suppose your billing cycle ends on the 28th of each month, and you are just shy of the 20MB limit on a metered plan when you switch to an unlimited plan on the 23rd of the month. You might think that you can use as much data as you want between the 23rd and 28th, but be sure to ask—when it comes to cellular billing, nothing is as simple as it appears.

—Brian Jepson

Discover Your Phone's Essential Numbers

Get the model number, serial number, and firmware version number of your smartphone. You will need them in future hacks.

In "Choose the Right Phone for the Network" [Hack #1] and "Pick the Right Class of Nokia Phone" [Hack #2], I discussed how to map Nokia smartphone models to wireless networks and Nokia device series. But what if you already own a Nokia phone? How do you determine its model number? How do you find out other crucial information about your phone that is useful in more advanced hacks? Well, read on.

The Model Number

Nokia smartphones are identified by cryptic model numbers, such as Nokia 6600, 6230, 3650, etc. Do you know which phone you own? Well, it's actually easy to find out your model number, even if you lost the original packaging box and order receipt. First, power off your phone, and then open the back cover of the device and remove the battery. You can read the Nokia model number off the white label (see Figure 1-5). Make a note of this number, as you will need to refer to it when applying hacks discussed throughout this book.

As I discussed in the beginning of this chapter, if you Google for your Nokia phone's model number (e.g., searching *Nokia 6600*), you can quickly jump to Nokia's official product page, which contains detailed technical information, related software downloads, and accessories for the phone.

The IMEI and Software Version Numbers

The International Mobile Equipment Identity (IMEI) code is a 15-digit number that uniquely identifies your mobile phone. It is useful for many purposes. For example, the wireless operator can associate your IMEI code with your account and identify stolen phones, locking them out of the network [Hack #24]. Third-party software on Nokia Series 60 devices often uses the

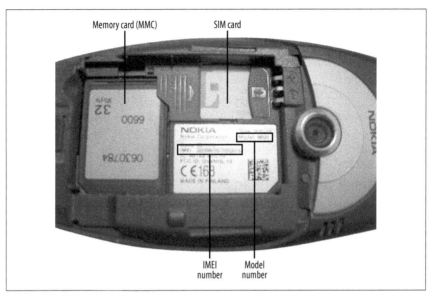

Figure 1-5. *The white label under the battery in a Nokia 6600*

IMEI code to tie the software license to the specific device it is installed on **[Hack #17]**. There are two ways to discover your phone's IMEI number:

- The first method is to look it up on the phone's label. The IMEI number should appear on the phone's packaging box. If you lost the original box, you can find it on the white sticker underneath the battery (see Figure 1-5).
- The second method is to type XS*#06# from your keypad as if you are dialing a phone number (do not press the green Send key). Immediately after you type those digits, the phone should display the IMEI number.

 If you purchase a phone from a third-party dealer, it is a good idea to make sure the IMEI number returned by the phone's software (the second method) matches the IMEI number on the sticker. A mismatch might indicate that the phone is stolen or counterfeit.

Another useful piece of information about your phone is its software version number. Devices with the same model number (i.e., the same hardware) but different software versions have slightly different behaviors. Some of the advanced hacks and applications depend on the device software version. Many of the technical bulletins published by Forum Nokia identify specific problems and bugs with specific software version numbers. To obtain the version number of your phone, enter the code *#0000# from your keypad on the idle screen, as though you are dialing a telephone number (see "Query and Change Service Settings" [Hack #8] for some other special operation codes). If your phone has an outdated firmware version, you can upgrade the firmware by "reflashing" the phone's main memory. You can send the phone to a service shop [Hack #7] to perform such upgrades.

> Reflashing your phone's firmware can harm the device in some cases. It is not covered under your regular warranty, so do it at your own risk!

Check Account Usage
#6   Keep track of your account to avoid expensive charges for additional minutes!

Today's wireless service contracts are designed by the operator and for the operator. As a consumer, you pay a predetermined monthly fee to subscribe to a service plan. This plan usually includes a certain number of voice call minutes and SMS/MMS messages, and a certain amount of data bandwidth. Calls made at different times (peak or off-peak), from different places (home or roaming areas), and to different destinations (family members, in-network mobile numbers, etc.) are all counted differently in the plan. If you go over the usage limits of the plan, you will be charged a hefty fee. Intimidated by such additional fees, many mobile subscribers choose plans that are more expensive than they need to make sure they never exceed the limit. The unused part of the service plan goes wasted at the end of each billing cycle. That's a lot of extra profit for the operator, at the subscriber's expense!

> Some U.S. wireless network operators offer solutions to the problem of unused minutes. On its Rollover Minutes plans, Cingular will allow you to carry over unused minutes from one month to the next. Sprint offers a different solution: its Fair and Flexible service plans will automatically switch to the next highest plan as your usage exceeds your limit.

As a well-informed power user, you can do better! You can beat the system and save money by tracking your account usage regularly and adjusting your usage patterns accordingly. For example, if you are running out of peak-time minutes near the end of a billing cycle, you can consciously make less important calls during the off-peak time, or even postpone them to the next billing cycle. By planning aggressively, you can reduce your month-to-month usage fluctuations. Such improved predictability allows you to choose a cheaper plan that closely matches your expected usage. The key to such successful planning is to gather accurate and real-time information on your account usage at any time. In this hack, I discuss how to do that.

Built-in Usage Log

Nokia Series 40 and Series 60 devices ship with a small program to keep track of phone use. On Series 60 devices, the program is called Log, and it is accessible from the Main menu. On Series 40 devices, the same function is accessible from the Call Registry menu. The Log program displays the total number of minutes for dialed and received calls, as well as the GPRS data bandwidth you've used. You can reset the log at the beginning of each billing cycle and use it to track account usage.

Of course, the problem with this approach is that the Log program is too simple to track today's complex service plans. The built-in log does not distinguish calls made at peak or off-peak times. It counts free calls, such as service calls (e.g., 611 or 911 calls), family plan calls, and in-network mobile-to-mobile calls, in the same way it does regular phone calls. It starts counting the minutes as soon as you press the Call button, regardless of whether the other party picks up the phone. In addition, the built-in log does not keep track of SMS/MMS messages.

Minutes Tracking Software for Series 60 Devices

If you use a Nokia Series 60 phone, you can replace the built-in Log program with one of several commercially available add-on programs. The one I like most is Moov Software's Minutes Manager. At the time of this writing, it cost $14.95 (you can download a free trial from its web site, *http://moovsoftware.com/mm_home.htm*). Minutes Manager is a Symbian OS-based program [Hack #17].

Minutes Manager tracks airtime minutes, data bandwidth, and messaging usage. You can configure it to support the most sophisticated service plans. For example, you can configure it according to the different off-peak and weekend time definitions used by different operators. It can start counting minutes either after the phone rings or after the other party picks it up. It can ignore certain free numbers specified in your contract (e.g., family numbers, 611 and 911 calls). It even handles small but important details such as whether peak minutes can be carried over to off-peak time and whether the plan permits minutes to carry over from the last billing cycle (e.g., the Cingular Rollover Minutes plans). Figure 1-6 shows the configuration options in Minutes Manager.

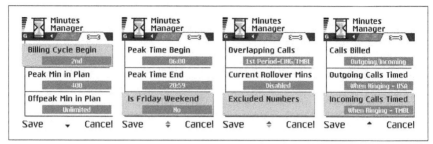

Figure 1-6. Configuration options in Minutes Manager

Minutes Manager resets itself at the beginning of each billing cycle. It displays the unused minutes left in the current billing cycle, both inside the program and on the phone's idle screen. The latter is a very convenient feature that helps you instantly know your available minutes, without pressing a key. Figure 1-7 shows the display information inside the Minutes Manager program.

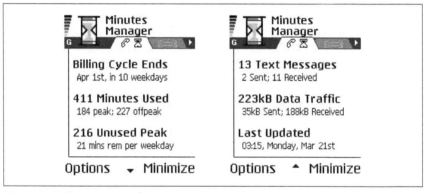

Figure 1-7. Minutes usage information in Minutes Manager

 Extended Log is another Symbian utility that analyzes the phone's call log and helps to track account usage. You can find more information about Extended Log, as well as purchase it, from its web site, *http://www.psiloc.com/index. html?id=165.*

However, Minutes Manager has some limitations. For instance, it is available only on Series 60 devices. Also, it cannot distinguish regular calls from in-network mobile-to-mobile calls, which are free in many plans.

Special Service Numbers

The best way to get accurate and authoritative account usage information is via your phone service's operator. Some operators allow subscribers to query current account usage via special service numbers or SMS messages. Here are some examples for the operators in the U.S.:

- If you are a T-Mobile subscriber, you can dial the special service number #646# (#MIN#) to check account usage (see Figure 1-8). T-Mobile also supports service number #225# (#BAL#) to check current account balances.

- If you are a Sprint PCS subscriber, you can dial the service number *4 for current account usage.

- If you are a Cingular subscriber, the service number #2455 (#BILL) is available in some markets for checking account usage.

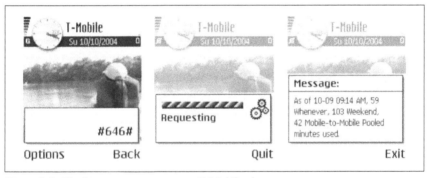

Figure 1-8. Checking account usage on a T-Mobile phone

 It might take a while before these balances are updated with calls made while you are roaming. This can be a very frustrating experience if you are on an extended trip in an area where your network operator relies entirely on roaming partners for its network infrastructure.

The SMS-based query involves sending an empty message to a special service number and getting the account usage via the response message. It is more popular in Europe and Asia than in the U.S. Most of these service numbers are accessible free of charge.

While the operator's service numbers provide convenient access to accurate account usage information, not all operators offer such services. The information provided is also very brief, due to the limited size of the response message. Your most up-to-date source of information will be your wireless network operator's customer service representatives.

Operator's Web Site

Most operators allow subscribers to monitor account usage via their web sites. To do this, you need to sign up for a free user account on the site. Once you've logged into the account, you can pay bills, change addresses, change service plans, and check your account usage. The information provided on the web site is typically more detailed than that provided via the service messages (see Figure 1-9 for an example from T-Mobile).

My Current Activity for (512) 300-6828 current as of: 10/08/04 06:21:30 PM

Summary of Unbilled Services			(Start Date: 10/02/04 - End Date: 11/01/04)		
Feature	Time Period ?	Type	Included	Used	Remaining
Call Forward Conditional	Whenever	Minutes	500.0	0.0	500.0
Free Minutes	Whenever	Minutes	400.0	59.0	341.0
Included Picture Messages	Whenever	Messages	Unlimited	0.0	Unlimited
Use Them Or Lose Them	Week End	Minutes	Unlimited	103.0	Unlimited
T-Mobile to T-Mobile	Whenever	Minutes	Unlimited	42.0	Unlimited
Use Them Or Lose Them	Off Peak	Minutes	Unlimited	0.0	Unlimited

Figure 1-9. Account usage from the T-Mobile web site

If you need to check your usage while you're on the go, you can use a web browser on the phone [Hack #50] to access the web site. However, most of those pages are not friendly to small screens. Figure 1-10 shows an example of how the nicely formatted table in Figure 1-9 is displayed via the Opera browser on a Series 60 device. To read the account usage from that page, first locate groups of three consecutive numbers (e.g., 400.0 59.0 341.0 on the third and forth lines in the second browser screen). Note that you should treat the word *Unlimited* as a number here as well. The first number is the plan limit, the second is the current usage, and the third is the remaining quota. Then, figure out the service item associated with the numbers from the text before these numbers. For example, from the text snippet "Free Minutes Whenever Minutes 400.0 59.0 341.0," I know I have used 59 minutes out of the 400 anytime free minutes in the plan. As you can see, checking usage from a browser on the phone is not very convenient—you'll be lucky if it gives you anything more than a headache. Plus, typing in username and password credentials on a phone keypad before you can log into your account can be a nuisance for most users.

Third-Party Alert Services

The operator-based methods share a common shortcoming: they are *pull-based*. That is, you have to actively query the operator to "pull" the account usage information. If you are very busy and forget to query the operator for a week, you might find yourself uncomfortably close to (if not already

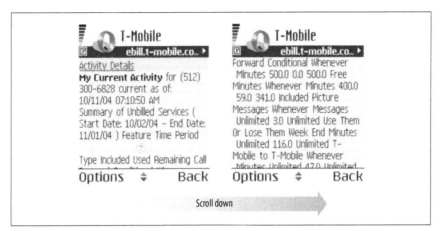

Figure 1-10. The T-Mobile account usage page displayed via the Opera browser on a Series 60 device

exceeding) your plan limits. A better approach is to get the information "pushed" to you as alerts when you reach certain points in your plan. The alerts can come in the form of SMS messages.

For $2 per month, an independent company named MinuteCheck can send you such account usage alerts. First you have to sign up for an account on the company's web site, *http://www.minutecheck.com/*, and provide your account information on your operator's web site. Once you grant MinuteCheck permission to log into the operator's web site on your behalf, the company can download your account usage page periodically and use an automated program to figure out the remaining minutes in your plan (a technique known as *screen scraping*). If you are close to your plan limit, the service sends you an SMS alert message. MinuteCheck supports subscribers of all U.S.-based operators.

AOL also provides a similar account usage alert service to its subscribers.

In this hack, I discussed several ways to track your account to avoid or reduce additional minutes charges. Each method has its advantages and disadvantages. You can choose one or a combination of these methods to suit your lifestyle. I use the Minutes Manager software, together with the operator's service number, to stay on top of my account usage.

Unlock Your Phone

#7   Unlock your phone so that it works with all network operators when you travel or switch plans. It is your phone. Take control of it!

If you bought your phone from a mobile operator, chances are very good that it's locked to the operator's network. While the phone itself might be compatible with all GSM networks (i.e., a triband phone), it rejects SIM cards from other GSM operators. Locking imposes some limitations on how you can use your phone:

- If you change your service plan to that of another operator, you will not be able to continue using the locked phone. You will have to buy a new phone and transfer all the data.

- The locked phone probably will not work with prepaid calling cards in foreign countries, since the prepaid service requires a new SIM card from a local operator. See "Use Prepaid Calling Cards" **[Hack #28]** for more on prepaid mobile phone calling cards.

- If you no longer use an old phone, it is difficult to sell or donate it if it is locked, as you have to require that the buyer or recipient has the same wireless operator as you do to make sure the phone functions as promised.

If you place a SIM card from another operator into a locked Nokia phone, it will display messages such as "Enter restriction code" or "SIM card rejected," and will refuse to even display the Main menu.

 If the error message is related to a SIM security code, or if it reads something like "SIM registration failed," you might not be able to connect to the network, but the phone is indeed unlocked. The acid test is to see whether you can access the Main menu. If you can, the phone is unlocked; if you can't, the phone is locked.

The rationale for operator locking is that the operator subsidizes the phone's price via service charges. So, it should do something to prevent the user from switching to other operators' services with the subsidized phone. However, from the customer's point of view, this is not a convincing argument, as the subsidy should be considered part of the marketing and customer acquisition cost. In fact, in most mobile service agreements, nothing prohibits you from unlocking the phone (check your service agreement to make sure this applies to you). In addition, the operator has no way of knowing whether your phone is locked.

As the legitimate owner of the phone, why should you put up with those limitations? In this hack, you will learn how to unlock and take control of your phone. An unlocked triband Nokia GSM phone will work with almost any GSM operator in the world. It is *your* phone, after all.

> If you bought your phone from an electronics store without a specific mobile service plan, it is probably unlocked and will work with all operators.

Just Call the Operator

Before you try anything yourself, you should call the mobile service operator's customer service department to see whether they are willing to unlock the phone for you for free. Some operators will unlock your phone under some circumstances. For example, T-Mobile USA might help you unlock your phone when you switch out of its service after the original service plan has expired. But the operator might charge an expensive fee ($40 or so for each phone) for such an unlocking service. In any case, call customer service and ask for a free unlock code—it never hurts to try.

If the operator agrees to unlock your phone, the customer service representative will give you a "remote unlock code" over the phone or via SMS. The code consists of numerical digits, the letters p and w, and the + sign.

You need to write down the code very carefully. Then, take the SIM out of the phone, turn on the phone, and enter the code. Press the * key four times within one second to get the letter w; three times within one second to get the letter p, and twice within one second to get the + sign.

If the unlocking is successful, the phone pops up a message indicating it has been unlocked (e.g., "phone restriction off"). You can put the SIM back in the phone and then power up, and you'll be back in business!

> It is very important to stress that you need to verify the unlock code and enter it very carefully. If you make a mistake, you can reboot the phone and try it again. But if you make a mistake five times in a row, the phone will not allow you additional attempts. You'll have to take the phone to a repair shop to get it unlocked in this case.

The operator-assisted unlocking service is easy and reliable. However, not all operators provide such services, and even those that do probably charge an expensive service fee. Of course, better and free (or very low-cost) alternatives are available from mobile hackers and third-party vendors. You will learn all about such solutions in the rest of this hack.

Calculate the Code Yourself

To calculate the unlock code, you need to download a code calculator program. Many code calculators are available as freeware, and others are for-pay shareware or commercial software. In general, the freeware calculators have a less-polished UI and require you to know more technical jargon.

> Calculating unlock codes using freeware calculators can be frustrating. If you do not enjoy searching the Internet and experimenting with software tools, I highly recommend that you use one of the low-cost unlock code services listed in Table 1-3, later in this chapter. This section is provided primarily for educational purposes.

You can find a comprehensive list of unlock code calculator programs from this web page: *http://www.unlockme.co.uk/downloads.html*. Not all calculators run on all flavors of Windows operating systems, and not all of them can calculate unlock codes for all Nokia models and all operators. So, you might have to download and experiment with multiple calculators. I recommend you try the CyberGSM, Hollowman, DCT4NCK, and NokiaFREE calculators listed on preceding sites.

> NokiaFREE has an online calculator at *http://unlock.nokiafree.org/*, which does the same thing as the Windows client software.

If the link on the web site goes out-of-date, you can simply search for the calculator name on Google and you will probably find a link to it. But for most freeware programs downloaded off an unknown web site, you need to be very careful with security. You should check the download with antivirus software and run it only in nonadministrator mode. Or better yet, run the calculator in an emulated Windows environment. Popular PC emulators are Virtual PC, VMware, and the excellent and free QEMU (*http://fabrice.bellard.free.fr/qemu*). The PC emulators also allow you to run those Windows-based unlock code calculators on Mac and Linux computers.

> Few freeware calculators can compute unlock codes for older Nokia models with the DCT3 firmware (NokiaFREE does, however). Most of today's Series 40 and Series 60 phones use the DCT4 firmware, which almost all calculator programs support. You can see a list of DCT3 phones at *http://www.unlockme.co.uk/softwareversions.html*. If your phone is not listed in the DCT3 table, it is probably a DCT4 phone.

Now, let's go through the calculation process using a basic (and early) freeware unlock code calculator—the DOS/Windows command-line utility, *DCT4NCK* (available from *http://homepage.ntlworld.com/danluik/downloads/dct4nck.zip*). Figure 1-11 shows *DCT4NCK* running under Windows 95 in QEMU. All the other unlock code calculators work similarly.

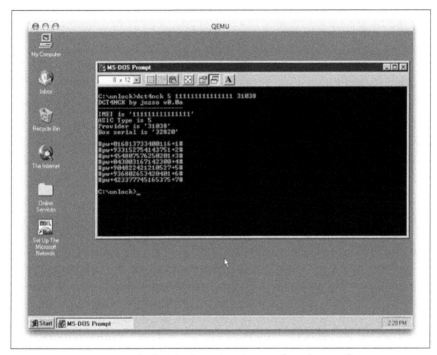

Figure 1-11. Generating unlock codes from the relative safety of an emulator

Once you run the dct4nck command from a DOS (shell) window, you will see the following usage message:

```
DCT4NCK by jozso v0.0a
----------------------
Usage: dct4nck locktype imei provider [boxserial]
```

The locktype, according to *http://unlockme.co.uk*, is usually 2, unless you're on a 3650 or 7650, in which case it's 5. The IMEI is the hardware serial number of your phone [Hack #5]. Finally, you'll need the provider code, which is a number you can pull from a network provider code list such as the one at *http://www.yeldar.co.uk/MCC-MNC.htm*.

The calculator generates seven different codes, at least one of which should unlock the phone when you enter it with the SIM removed, but you get only five tries before the phone locks *you* out. So, you really want to double-check your work before entering would-be unlock codes into the phone.

Typically, you want to try the fourth one first, then the first one, and then the seventh one.

Although the unlock code calculation process might sound easy, it has several caveats, especially with freeware calculators. For instance, the following notes about AT&T Wireless's Nokia 3650 are available on the *http:// unlockme.co.uk/* web site:

- Although a provider code for AT&T is listed in the network provider list at *http://www.yeldar.co.uk/MCC-MNC.htm*, Nokia 3650 users on AT&T Wireless have a variety of codes to choose from depending on their phone's IMEI.

- For Nokia 3650 on AT&T USA, always use the first code out of the seven generated.

The notes and exceptions will continue to increase as new phone models are released and operators find more ways to make unlocking more difficult. Your best bet is to take the time to read *everything* you can find at *http:// unlockme.co.uk*.

Of course, it is possible to keep track of those exceptions and integrate them into the logic of the unlock calculator software. For instance, the program could just ask you the phone model, IMEI, and operator name. If you enter Nokia 3650 and AT&T, it should be able to figure out the correct operator code automatically, and it should return only the first unlock code. But most freeware authors do not have the time for such extensive research and do not tend to keep their program updated after its release. You have to live with those inconveniences—you're getting the software for free, after all.

Get Professional Help

By now, you probably have concluded that you do not want to go through the hassle of calculating the unlock code yourself, for the following reasons:

- Running untrusted Windows software is never a very good idea.

- If you miss one rule exception, you will not get the correct unlock code.

- Even if you do get the calculator working correctly, you typically get seven codes, and the phone allows only five attempts before it locks itself up.

The good news is that you do not need to calculate the unlock code yourself or beg the operator for it. For a couple of dollars, you can get an expert to calculate it for you using the most up-to-date software. Then you can receive the unlock code instantaneously via phone or the Web. For most users, that is the best way to unlock a phone. Table 1-3 shows a list of popular unlocking services. All prices are current as of April 2005.

Table 1-3. Popular unlock code services

Web site	Price	Service	Coverage
http://www.dial-a-code.co.uk/	One British pound per minute, with each case lasting about five minutes	Phone/voice	All Nokia devices on any UK network
http://www.freeyourphones.com/	$2.99	Email	Operators around the world
http://www.unlock123.com/	$4.95	Web	Operators in North America and the Philippines
http://www.mobilefun.co.uk/	$2.95	Web	Some selected UK and U.S. operators
http://www.mobileliberation.com/	$2.95	Web	Operators around the world
http://www.lockfree.com/	$2.99	Web	Operators around the world

Once you get the unlock code, you can enter it into the phone using the remote unlocking instructions discussed earlier in this hack. Or, you can simply follow the instructions provided by the unlocking service.

Use a Cable-Based Solution

An alternative solution to remote locking is to unlock the phone using data cables. The cable-based solution is your only choice if you failed to remotely unlock the phone after five attempts. (As noted earlier, the phone automatically locks itself after five attempts.)

Cable-based solutions can be complex. And if you do not know what you are doing, you can permanently damage your phone. Hence, if you are a casual mobile phone user, I highly recommend you find a professional mobile phone service shop to do the unlocking. These shops have special devices known as *unlocking clips*, which are preloaded with the latest commercial unlocking software. All they have to do is just hook the unlocking clip with the phone via a cable and press a button, and the phone is unlocked. These professionals can also reflash the entire phone firmware and return it to its factory setting via a cable connection. This way, you can get rid of any security code and clear past failed unlock attempts.

> Of course, you can buy an unlocking clip yourself, but it is expensive, so it's not economically feasible if you're buying it to unlock only a couple of devices.

To do cable-based unlocking, you need to purchase a serial or USB cable for your phone from web sites such as *http://ucables.com/cables/Nokia*. You need MBus interface cables to work with most unlocking software. Those cables cost around $50 each. After hooking up the phone with the computer, you can use software such as JIC DCT4 Unlocker (*http://homepage.ntlworld.com/danluik/downloads/JIC%20dct4%20unlocker.zip*) to unlock the phone directly from the computer. Some programs also allow you to reset the security codes, etc., for some phone models. You can read more about various cable-based unlocking solutions at *http://unlockme.co.uk/*.

> You also can use the data cable with the Nokia PC Suite to transfer data between the phone and the computer **[Hack #15]**.

This hack is based in part on an O'Reilly Network article (*http://www.oreillynet.com/pub/wlg/3935*) written by Schuyler Erle.

—Michael Yuan

HACK #8   Query and Change Service Settings

You can dial numbers on your keypad to get billing information from your cellular operator or to configure forwarding settings.

If you're a T-Mobile USA subscriber, you can dial #646# (#MIN#) **[Hack #6]** and a message will come back that tells you what your minute usage is for the billing period. You also can dial other little features on your phone to get interesting information or change network settings such as call forwarding.

On GSM networks, you can check the status of many services by sending messages from your handset to the operator's network. Some GSM networks, such as T-Mobile USA, offer # codes, as shown in Table 1-4. You need to press the Call button (the green phone button you press after dialing a number) after you type in these numbers. Short * codes work in the same way—some of these are shown in Table 1-5. You should check your phone manual or your operator's web site to learn which codes you can use.

Table 1-4. Short # codes

Short code	Result	Example operator
#MIN# (#646#)	Sends the minutes usage to your handset as a message	T-Mobile USA
#NUM# (#686#)	Sends your mobile number to you as a message	T-Mobile USA

Table 1-4. Short # codes (continued)

Short code	Result	Example operator
#999#	Sends the balance of your prepaid account to you as a message	T-Mobile USA
#BAL# (#225#)	Sends your balance to you as a message	T-Mobile USA

*Table 1-5. Short * codes*

Short code	Result	Example operator
*MIN# (*646#)	Sends your minutes usage to your phone as a message	Cingular
*BAL# (*225#)	Gives you an update on your outstanding balance	Cingular
*NUM (*686)	Reads your mobile number to you with a voice recording	T-Mobile USA
*PAY (*729)	Enables you to make a free phone call to pay your bill	Cingular
*#100#	Sends you your mobile number as a message	Vodafone
*#1345#	Sends you your balance as a message	Vodafone
*#103#	Displays network time and date	Vodafone
*#147#	Displays the last incoming call, including time and date	Vodafone
*#105#	Displays cell tower information	Vodafone

You can also change your call preferences. Table 1-6 shows some call codes for GSM networks. (Forwarding may incur extra charges.)

Table 1-6. Call codes for GSM networks

Command string	Result of command
#31#PhoneNumber	Masks your phone number from Caller ID (your call to PhoneNumber shows up as "Private Call").
*#21#	Checks status of call forwarding.
**21*PhoneNumber#	Forwards all calls to PhoneNumber.
##21#	Resets call forwarding to default.
*#67#	Checks status of forwarding for calls that arrive when you are busy.
**67*PhoneNumber#	Forwards all busy calls to PhoneNumber.
##67#	Resets busy call forwarding to default.
*#61#	Checks status of forwarding for calls that you don't answer.
**61*PhoneNumber#	Forwards all unanswered calls to PhoneNumber.

Table 1-6. Call codes for GSM networks (continued)

Command string	Result of command
##61#	Resets unanswered call forwarding to default.
*#62#	Checks status of forwarding for calls that arrive when you are out of reach (when your phone is off or out of a service area).
**62*PhoneNumber#	Forwards all unreachable calls to PhoneNumber.
##62#	Resets unreachable call forwarding to default.

Although many Nokia phones have menu options for the forwarding settings, some of them are hard to get to (for example, most Nokia 3650s have a Forwarding menu, but it is hidden—to access it, you must assign one of your soft keys to it). And if you ever have to configure forwarding settings from a phone that doesn't have an option to configure them, you can pop in your SIM and use these codes (forwarding settings are per-subscriber, not per-phone).

My friend, Jesse Chan-Norris (*http://www.jessechannorris.com/*), pointed out to me that Verizon Wireless has similar codes; these are shown in Table 1-7. You need to press the Call button after you type in these numbers.

Table 1-7. Verizon Wireless short codes

Short Code	Result
#MIN (#646)	Sends the minutes usage to your handset as a text message or voice message
#BAL (#225)	Gives account balance information as a text message or voice message
#PMT (#768)	Enables you to make a free phone call to pay your bill
*DWI (*394)	Reports a suspected drunk driver (might not work in all service areas)

—Emory Lundberg

Extend Your Talk Time
#9   Avoid battery drain by conserving power, using mobile chargers and spare batteries, and practicing proper battery maintenance procedures.

Your Nokia phone displays a battery gauge on the right side of the idle screen to indicate the remaining battery charge. The battery gradually loses its charge even if the phone is powered off. Therefore, it is essential to recharge the battery every several days to keep the phone in a ready-to-use state. Older batteries need recharging more often than new ones, since the battery loses its capabilities as time goes on.

It is helpful to understand how batteries are charged. In most cases, both the battery and mobile phone contain electronics that monitor and manage the charging cycle. A full battery charge occurs in two phases. The first is the quick charge, which generally takes the battery up to an 80% charge. The second is the trickle charge, which slowly tops off the battery. As the battery gets older, it holds a smaller charge, so the time required to get to a full charge will vary.

If the battery charge is too low, the phone will alert you periodically via a "low battery" message accompanied by a single beep. You can silence the beep by turning the phone to a Silent profile. From the first "low battery" message, you have about an hour of standby time or 10 minutes of talk time until the battery finally runs out. A mobile phone with a drained battery can cause lost productivity, missed opportunities, and even anxiety among some people. To avoid battery drain, you can proactively adjust your usage pattern to conserve power and keep a second battery on hand as a backup. And, for the longer term, proper battery maintenance practices help prolong battery life and improve the capability of older batteries. In this hack, I cover all these topics.

Conserve the Battery Power

First, try to make voice calls only from places where the signal strength is strong. This way, the radio doesn't need to amplify its signal, and hence, you save energy. You can also set up the phone to make GPRS connections only "when needed" as opposed to "when available" to save the extra energy needed to maintain an always-on GPRS connection. You can change the GPRS connection setting via the Tools → Settings → Connection → GPRS menu on a Series 60 device, and via the Settings → Connectivity → GPRS menu on a Series 40 device (see Figure 1-12).

In addition, you can conserve battery power by reducing the use of network-intensive applications such as the web browser, the MMS client, and the email client. Those applications need to make frequent GPRS connections to the network.

The backlit LCD is probably the biggest battery hog on the phone. You can reduce the brightness and increase the contrast of the LCD to conserve energy. You can also reduce the screensaver time-out to reduce the length of time the backlight is turned on. On a Series 60 device, you can adjust the LCD backlight brightness and screensaver time-out via the Tools → Settings

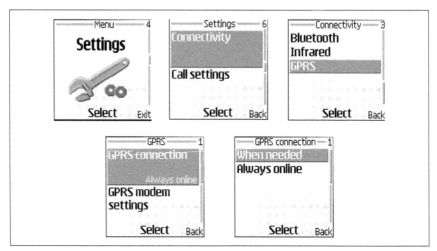

Figure 1-12. Adjusting the GPRS connection setting on a Series 40 device

→ Phone → Display menu (see Figure 1-13). Some Series 40 devices support adjusting the LCD brightness via the Settings → Display settings → Display brightness menu. You can avoid accidentally turning on the LCD backlight by locking the keypad when the phone is not in use.

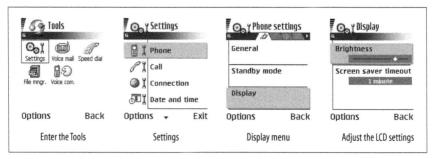

Figure 1-13. Adjusting the LCD settings to conserve the battery

On the other end of the spectrum, a Symbian program called PowerTorch allows you to turn on the backlight for an extended period of time so that you can use the phone as a torch (or flashlight). The program runs in the background and is activated through a hotkey. You can adjust the brightness and color of the light using the phone's scroll navigation key. You can purchase PowerTorch from *http://www. handango.com* and many other Symbian software stores. A similar freeware package called Torch is also available. You will be amazed at how well a cell phone can work as a flashlight.

Since the Bluetooth radio consumes battery power whenever it is turned on, regardless of whether it is actually connected to a network, you should turn on Bluetooth only when you are ready to connect to other devices. In general, you should not leave the Bluetooth radio turned on when the phone is idle. You can turn Bluetooth on and off from the Connect → Bluetooth menu on a Series 60 device, and from the Settings → Connectivity → Bluetooth menu on a Series 40 device. Once Bluetooth is turned on, you should see a solid black ball on the screen. If there is an active Bluetooth connection to or from the device, you should see braces around the black ball (see Figure 1-14).

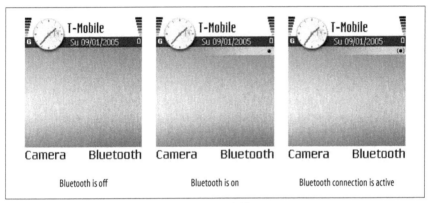

Figure 1-14. The Bluetooth status indicator available on-screen

Nokia Series 60 devices support multiple concurrent applications [Hack #19]. However, too many background applications could slow down your phone and drain the battery. In addition, you should always properly exit all applications before you turn off the phone. If you turn off the phone with applications running in the background, the phone might have to perform some very battery-intensive operations to forcefully terminate the running threads, instead of closing applications gracefully.

You can extend your phone's battery life by avoiding mobile games. Mobile games, especially 3D action games, make heavy use of the CPU and LCD. If you have to play mobile games, you can conserve energy by turning off the speakerphone, the vibrator, and the LCD backlight. You can often do that via the game's menu and/or by selecting the Silent profile [Hack #43].

The last tip for conserving battery power is to reduce the use of multimedia applications. The phone camera requires a lot of current to operate the shutter, the camera sensor itself, and the virtual viewfinder on the LCD. For media players, decoding MP3 audio and MPEG4 video (in 3GPP video clips) is very CPU intensive, and CPU usage is proportional to power draw.

Mobile Chargers

Energy conservation has its limits. You still have to use the phone, after all. If you are a frequent traveler or commuter, you might have to recharge your battery on the go. Let's explore some options here.

If you are a car traveler, you can purchase a car charger for your phone. The car charger uses the 12V DC power outlet in the cigarette lighter to charge the phone battery. The same cigarette lighter outlet is also available on some airplane seats. Nokia car chargers are available anywhere Nokia phones are sold. Since different battery models require different chargers, make sure your charger is compatible with your phone and your battery. Alternatively, you can purchase a power inverter to convert the cigarette lighter outlet to a regular 110V/220V AC wall outlet. Then you can use the regular wall charger that comes with the phone. Power inverters are available in many electronics stores, such as Best Buy and Circuit City in the U.S.

> Do not keep your car charger or power inverter plugged into the car for an extended period of time if the engine is not running, since it might drain the car battery.

If you carry a laptop computer most of the time and you need an emergency charger in case the battery runs out, you can purchase a USB-based charger. It plugs into the computer's USB port at one end and the phone at the other end. It uses the larger battery in the laptop to charge the smaller battery in the phone. Since the USB charger is lighter and smaller than the regular wall charger, it is an excellent replacement charger for home and office as well. An example of such a USB-based mobile charger is the ZIP-LINQ (*http://www.ziplinq.com*) retractable charger cables and voltage boosters.

If you are a train or bus commuter, you might be able to find coin-operated charger stations at train or bus stations, especially in Europe. You might have time to charge the phone for only 5–10 minutes, but that will give you a couple of hours of standby time and tens of minutes of talk time. That is probably enough to last until you get to the office or your home.

Another emergency backup battery solution is to use a nonrechargeable battery, which either replaces the internal battery or plugs into the phone's power plug. An example of such a product is the Cellboost instant recharger (available from the *http://www.cellboost.com* web site and in many retail locations). The benefit of this approach is that the nonrechargeable battery can be made to have a large capacity and, more importantly, a very long shelf life. You can carry it around for years until you experience a real emergency.

Spare and Replacement Batteries

For real heavy users, and for those who seek peace of mind, I suggest buying at least one spare battery. You can carry a fully charged backup battery with you in an antistatic plastic bag. Since the backup battery loses its charge in about 1–2 weeks (the shelf life), you must rotate the batteries to keep both of them fresh: once the first battery is drained, replace it with the second one. But when you get home, fully charge the first battery and make it the new backup battery.

Nokia sells original batteries for its devices through Nokia authorized resellers. Those batteries are of high quality and conform to the highest safety standards. However, they are also relatively expensive. You can often buy cheaper Nokia batteries from unauthorized dealers or via the Internet. However, those batteries could come from third-party manufacturers and their quality is not consistent. You could get very good-quality or poor-quality batteries from those vendors. So, make sure you obtain a warranty with your purchase. It is easy to distinguish third-party batteries from Nokia original batteries since third-party vendors are not allowed to use the Nokia brand name. The labels on third-party batteries typically contain phrases such as "Replacement of" or "Compatible with." They often use "NK" or "NOK" to refer to Nokia model numbers.

Then, there are counterfeit Nokia batteries, which falsely claim to be original Nokia batteries. You can identify counterfeit batteries from their crooked labels and packaging. The labels of counterfeit batteries often contain misspelled words or other obvious printing errors. Counterfeit batteries are illegal and could pose a fire hazard due to the lack of an internal fuse. You should not buy or use them. If you are at all unsure about the source of a battery, buy one from an authorized dealer.

Battery Maintenance

The rechargeable battery in a Nokia mobile phone is typically a NiMH battery. Unlike older generations of rechargeable batteries (e.g., NiCd batteries), the NiMH batteries do not suffer from the "memory effect." That means you can recharge the battery anytime you want. This helps you to keep the mobile phone ready at all times. However, since batteries are designed to have 500–1,000 recharging cycles, I do not advise that you recharge the battery too frequently. Typically, recharging once a day or once every two days is a good compromise. For the same reason, you should not connect the phone to the charger for an extended period of time (e.g., several days). Since the battery loses its charge naturally, frequent recharging cycles occur when the charger is always connected.

The biggest factor that shortens the life of NiMH batteries is high tempera-ture. You should not leave the phone or the backup battery in a hot car or in direct sun. This is another reason excessive recharging is bad—the battery heats up as the recharging current passes through it.

Battery Disposal

Under no circumstances should you ever dispose of any rechargeable bat-tery in your trash. Rechargeable batteries contain highly toxic materials that can harm the environment and jeopardize the health of waste management workers. In the U.S., you generally can dispose of batteries at any store that sells electronics—this ensures that the battery will be recycled or disposed of safely. Best Buy, an electronics superstore chain, has battery disposal bins in the front lobby of each store. They also have bins for old cellular phones and ink cartridges.

If you are disposing of a phone along with a battery, one option is to con-tact your local police department to see if they have a partnership with a local women's shelter that takes donations of mobile phones. Many shelters loan out unsubscribed handsets, which can still be used to dial 911, to women who need their services. Your old mobile phone could save a life!

## CHAPTER TWO

Get Connected
Hacks 10–15

By their very nature, mobile phones are connected devices. They are useful only when connected into the wireless network. Connecting a mobile phone to the wireless voice network is straightforward. Just turn on the phone and it automatically finds and registers on the network.

Smartphones, however, have features well beyond those available with simple mobile phones. To fully utilize the data-processing power of smartphones, you should connect them to the wireless data network. In this chapter, I cover the Nokia smartphone's data connection configuration, as well as issues with both the wireless Internet and local Bluetooth networks.

HACK
#10
Connect Your Phone to the Internet

Your mobile smartphone is as powerful as desktop computers were not too long ago. Not only that, but you can connect to the Internet with it. Combine these capabilities, and you've got a smart device that can connect you to email, the Web, and more.

The mobile Internet is a key part of the mobile lifestyle. Many of the cool features crammed into your Nokia phone are designed specially for Internet use. However, an October 2004 mobile survey conducted by Wacom Components suggests that more than half of the users polled thought that it was too difficult to access the Internet from their mobile phone, and hence they avoided the mobile Internet altogether.

If you're like these users and your phone is not yet connected to the Internet, you are not only missing out on a lot of fun, but you've also wasted your money by buying an expensive gadget that you don't use to its fullest extent!

Once you understand the key concepts involved, accessing the Internet from your Nokia phone becomes very easy. In this hack, I'll tell you all about it.

First I'll discuss the basic concepts of data access settings on a Nokia device. Then I'll cover how to determine the right setting values for your device and your network.

Data Access Settings

A Nokia device can access the Internet via several different data bearers [Hack #4], wireless operators, and proxy servers. Each valid combination of such parameters is known as a *data access setting*. A Nokia device can hold and manage multiple data access settings. You can assign a different setting for each application, or activate a different setting when you roam to a new wireless network. The Nokia Series 40 and Series 60 devices [Hack #2] manage data access settings in different manners.

On a Series 60 device, all data access settings are centrally managed as *access points*. You can define new access points or edit existing ones via the Tools → Settings → Connections → Access Points menu (see Figure 2-1). You can assign a default access point to each network-aware application on the device via the application's own settings menu. For the Services application (your phone's web browser [Hack #50]), the default access point is specified via Options → Settings → Default access point. For the Messaging application, the default access point for MMS is specified via Options → Settings → Multimedia Messaging → Access point in use, and the access points for email mailboxes [Hack #60] are specified via Options → Settings → E-mail → Mailboxes → (Name of your mailbox) → Access point in use. For custom installed Java and Symbian applications, the default access point for each application is specified in the application manager [Hack #16] and [Hack #17]. If a network application does not have a default access point, you will be prompted to choose an access point when the application tries to connect to the network (see Figure 2-2).

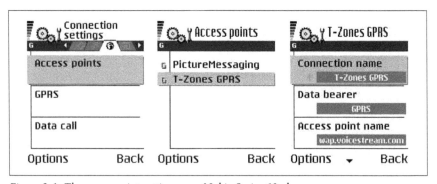

Figure 2-1. The access point settings on a Nokia Series 60 phone

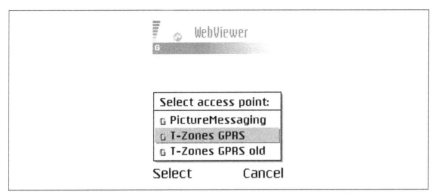

Figure 2-2. The device prompting you to choose an access point if no default access point is assigned for this application

> The term *access point* in this section refers to a valid combination of access parameters on Nokia Series 60 devices. Do not confuse it with the term *GPRS access point name*, discussed later in this hack.

On a Series 40 device, each application manages its own set of data access settings. For instance, the MMS data connection settings are available under the Messages → Message settings → Multimedia msgs. → Connection settings menu. The WAP and general Internet connection settings are available under the Services → Settings → Connection settings menu. The General Packet Radio Service (GPRS) modem connection settings are available under the Settings → Connectivity → GPRS → GPRS modem settings menu. Figure 2-3 shows how to list and edit the access settings in the Services application.

> The third-party Java games and applications on a Series 40 device share the connection settings from the Services application.

Now let's look at the available options in each data access setting (or connection setting) in more detail.

GSM dial-up modem. If the data bearer is GSM Data, you have to specify the phone number of the operator's dial-up modem. You can skip this step if you are using the GPRS data bearer.

When an application accesses the Internet via GSM Data, the phone automatically dials the modem number to establish the data connection. You need to specify the type of data call in the configuration. The analog call

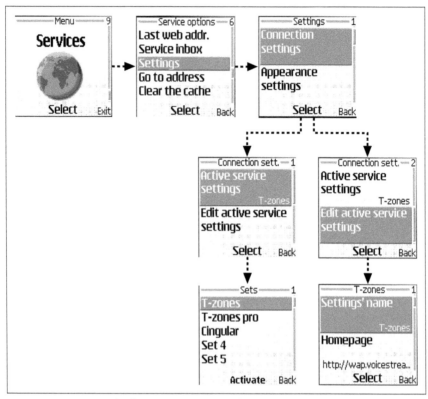

Figure 2-3. Managing the data access settings in the Services application on a Series 40 device

modulates data over analog voice signals, which can be a slow and unreliable process. Most operators today use an ISDN connection for data calls (although the speed is still very limited). Notice here that since you are charged the regular rate for this call, you should use a local number if possible. For example, if you travel to a foreign country and still use your home country's service number, you might be charged an international fee for each minute of the call.

The data call times out after a certain period of inactivity to avoid excessive airtime usage. The default timeout is five minutes. You can configure this value via the Tools → Settings → Connections → Data Call menu on a Series 60 device. However, you should disconnect the connection proactively to save airtime. If you use GSM Data to browse web pages, you should disconnect it as soon as you are done, or if you expect a long period of inactivity.

GPRS access point name. If you use GPRS as the data bearer, there is no modem number to dial because the wireless network is already packet-switched and the data connection is "always on." But the wireless data network and the TCP/IP-based general Internet locate destination addresses and transport data packets differently. The wireless protocol stacks (e.g., TCP/IP Wireless Profile) are optimized for low bandwidth and high reliability. The wireless operator provides the interface between the GPRS network and the general Internet using a server known as the GPRS access point name (APN). In the wireless network, the APN is identified by a string name. You have to enter the APN in the setting for GPRS data bearers. Notice that the GPRS APN is a different concept from the "access point" on the phone. The latter refers to a valid combination of access settings, including the name of the GPRS APN.

Some operators require the phone to use a username and password pair to authenticate itself to access the GPRS APN (see Figure 2-4). In most cases, this is just a very simple username and password pair that is the same for all subscribers.

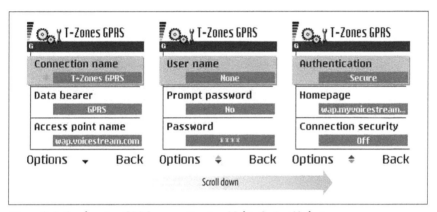

Figure 2-4. Configuring GPRS connection in a Nokia Series 60 device

Proxy servers. The GPRS APN allows generic data packets to pass from the wireless network to the Internet and vice versa. But for many applications, the high-level data protocols on the wireless side are also different from those on the Internet side. Proxy servers (or gateway servers) are used to interface those application-level protocols. Two important types of proxy servers for Nokia devices are the MMS and WAP proxies:

- The MMS proxy receives MMS messages from the phone and translates them to the standard Simple Mail Transport Protocol (SMTP) for Internet delivery. When an MMS message alert is delivered to a phone, the phone retrieves the message's multimedia content over the Internet via the MMS proxy.

- The WAP proxy (a.k.a. WAP gateway) fetches Wireless Markup Language (WML) pages on the Internet using the HTTP protocol. It preprocesses WML pages, compresses them, and resolves WMLScripts according to the WAP specification. Then the WAP proxy feeds the processed pages to the phone browser using the WAP Session Protocol (WSP) over the wireless network.

On a Series 60 device, you can configure the proxy settings for an access point by choosing the Options → Advanced settings menu from the access point configuration screen. You need to know the proxy server's IP address and service port number (see Figure 2-5 for MMS proxy settings on T-Mobile networks). Via the advanced settings menu, you can also enter a static IP address for your phone or DNS servers, if you are assigned one by the operator. The default values for the IP address and name servers are 0.0.0.0, which means that, once connected, the phone is to obtain those addresses automatically from the network.

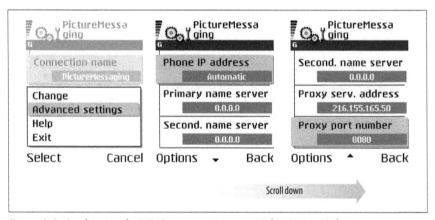

Figure 2-5. Configuring the MMS proxy setting on a Nokia Series 60 device

As I mentioned, the MMS and WAP settings are configured within the Messages and Services applications, respectively, on Series 40 devices. Figure 2-6 shows the configuration of a WAP proxy on a Nokia 6230 device. It is accessed via the Services → Settings → Connection settings → Edit active service settings menu. Select "Bearer settings" to change the APN and GPRS settings mentioned in the previous section.

Determine the Settings for Your Phone and Operator

Each wireless operator uses its own dial-up number, GPRS access point node, and proxy server configurations. If you purchased your phone directly from your wireless operator, you should have the appropriate Internet settings defined in the phone already. But in some cases, such as when buying a

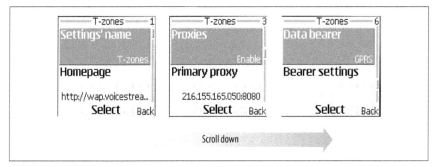

Figure 2-6. Configuring the WAP proxy setting on a Nokia Series 40 device

used phone or an unlocked phone, you will have to define the access points yourself. You can also call your wireless operator's customer service department to have them *provision* the phone—they will send Over The Air (OTA) messages to your phone that configure it automatically. You can also contact customer service to find out the exact settings you need in case you need to enter them manually (you will probably need to do this if you are using a phone that your wireless operator does not support).

 The Nokia 6230 supports only GPRS-based connections. You have to use the Settings → Connectivity → GPRS → GPRS modem menu settings to select the access points and GPRS APNs. To define a new access point, you can simply activate a blank access point and then edit the settings for the activated access point.

If your operator's customer support cannot help you with the settings (e.g., you have an "unsupported" phone!), you can search the Internet to find out what they are. For example, the web site *http://www.opera.com/products/ mobile/docs/connect/* lists GPRS APN names for many operators around the world; similarly, the web site *http://www.filesaveas.com/gprs.html* lists GPRS, WAP, and MMS settings for several operators in the United Kingdom.

Automatically Provision GPRS Settings

So far, I have covered the important concepts regarding access point settings. But finding the exact settings and then entering them by hand is still tedious and error prone. There is an easier way. You can use a web site maintained by Nokia to remotely configure your phone. The URL is *http:// www.nokiausa.com/support/settings/*. You are asked to choose your region, operator, phone model, and type of service setting (e.g., WAP, Email, or MMS) you want to provision to your phone. Figure 2-7 shows the web-based configuration interface. After you type in your phone number and click OK, the web site sends a WAP service message to your phone's message Inbox. If you open that message, the new connection setting is automatically entered into your Access Points menu. As discussed earlier in this hack, your wireless operator can also provision your phone (but it might be faster to use web-based provisioning).

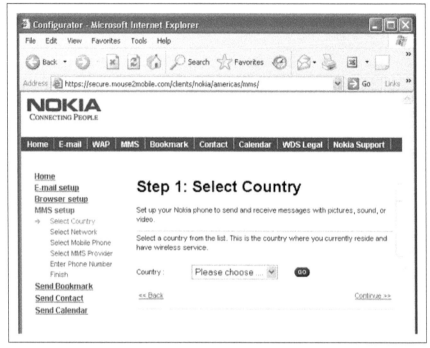

Figure 2-7. Using the Nokia web site to send MMS configuration settings as a short message to your phone

One limitation of this method is that you can only choose a phone model that is officially supported by your operator. If you have a new, unsupported phone, you can probably get away with choosing a similar older model. For example, settings for the Nokia 3650 phone work fine on the newer Nokia 6600 phone.

Share the Internet Connection from a Computer

If a Nokia Series 60 phone is connected to a computer via cable or Bluetooth [Hack #11], it can share the computer's network connection. With the computer acting as the proxy between the phone and the Internet, you can play with Internet applications on the phone without paying the GPRS subscription fee. However, the practical value of such a hack is questionable, as it limits the phone's mobility.

It is a complex process to set up both the correct routing tables on the computer and the correct access point on the phone. You have a better chance getting the computer proxy access point to work if you have working knowledge about TCP/IP networking.

Here are some resources and online tutorials that cover how to set up the computer network proxy for the phone:

- Mika Raento has an excellent tutorial covering Windows and Linux computers with Nokia 7650 and 6600 phones. It is available at *http://www.cs.helsinki.fi/u/mraento/symbian/bt-ap.html*.

- Rael Dornfest and James Duncan Davidson's book, *Mac OS X Panther Hacks* (O'Reilly, 2004), has a hack covering how to share a Mac computer's Internet connection to a Nokia Series 60 phone via Bluetooth. This hack is freely available in PDF format at *http://www.oreilly.com/catalog/0596007183/chapter/hack55.pdf*.

To avoid configuration headaches, you can buy a dedicated Bluetooth Internet access point device, such as PicoBlue, to provide Internet connections directly to Bluetooth devices. The Bluetooth Internet access point is preconfigured and replaces the proxy PC. It could save you a lot of trouble. But it is also pretty expensive ($500 or so).

A reverse of this hack is to use the phone's GPRS connection to provide Internet access to computers (e.g., your laptop while you are traveling). That is a much more useful hack and is covered in "Connect Your Computer to the Mobile Network" **[Hack #40]**.

Use Bluetooth to Replace Cables

We all use cables to connect things, but this can rapidly become unwieldy. Use Bluetooth to connect your phone to other phones, PDAs, computers, and more. Bluetooth makes it easy for mobile devices to become part of your personal network.

Bluetooth is a short-range (about 30 feet, or 10 meters) wireless technology that operates in the unregulated 2.4GHz radio band. It is developed by the Bluetooth Special Interest Group (SIG), which consists of more than 2,000 companies, including Nokia. Bluetooth is designed as a "cable replacement" technology to cut down on cable clutter in the increasingly digital home and office. For instance, Bluetooth-based keyboards and mice have gained a lot of popularity these days.

Your WiFi network and some cordless phone systems operate in the same 2.4GHz radio band. The interference
between those networks is generally negligible.

Bluetooth is ideally suited for mobile phones and handheld devices. Smartphones are not only voice communications tools but also tiny computers.
Like regular computers, mobile devices are most useful when they work in
collaboration with peer computers or devices. However, due to the mobile
nature of those devices, cable-based solutions are inconvenient. You
shouldn't have to carry around USB cables or cradles just to synchronize the
contact list from your phone to both your home and office PCs!

Bluetooth technology enables you to build ad hoc personal networks that
connect all nearby devices wherever you go. Since the mobile phone is
always with you and (usually) contains your most up-to-date personal information, it is a central component in the personal network. For example, the
phone could provide Internet access to any laptop computer you happen to
work with through your GPRS subscription [Hack #40]; the phone could
become the remote control for your computers [Hack #41] and [Hack #42]. Many
hacks in this book rely on Bluetooth to work.

Most Nokia devices are sold without the serial or USB
cables. If you opt for a cable-based solution, you will have to
purchase the cables separately. Also, you need special software to communicate with the device via cables. The Nokia
PC Suite is a good choice for such software on Windows PCs
[Hack #15].

Most new Nokia phones, especially those that came out after 2003, are Bluetooth enabled. You can simply navigate to the Connectivity → Bluetooth
menu to turn on the Bluetooth radio. You will see a solid black ball on the
phone's idle screen when Bluetooth is enabled. Remember to give your
device a unique Bluetooth name so that it can be identified on the network
(see Figure 2-8).

In this hack, I'll cover the basic concepts and operations related to Bluetooth networks. The detailed configurations for setting up a Bluetooth network between a Nokia smartphone and a Windows PC, Mac, or Linux
computer are covered in subsequent hacks.

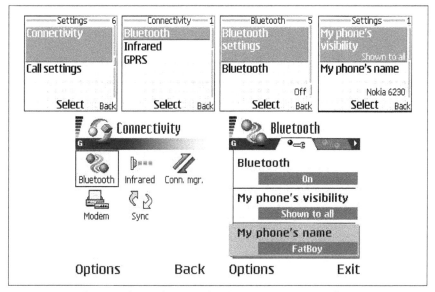

Figure 2-8. Configuring Bluetooth on a Nokia Series 40 or Series 60 phone

 Bluetooth applications go beyond cable replacement. The Nokia Sensor is a Bluetooth-based social networking application for Series 60 smartphones. You can download it for free from the Software section on the Nokia web site (or you can simply search Google using the phrase *Nokia Sensor* for a direct link). Using Nokia Sensor, you can publish a profile of yourself, including pictures, job titles, and hobbies, through Bluetooth. When other Nokia Sensor users are in your proximity, they can see your profile via Bluetooth. If they find you interesting, they can decide to approach you for conversation, send you business cards, leave a note in your "guest book," or send you messages. This application can be very useful in conference halls, cinemas, or bars. It allows you to find strangers with common interests.

Bluetooth Networks

A key characteristic of a Bluetooth network is its dynamic nature. Devices can join and leave the network at any time. Any device in the network can search all its peers and discover the network services they provide. A Bluetooth network needs a flexible and robust architecture.

The *piconet* forms the basic structure of a Bluetooth network. A piconet consists of a hub device, and one or more spoke devices that are connected to the hub. The hub device is known as the *master*, and each spoke is known as a *slave*. In addition:

- Any Bluetooth device can be a server or a client.
- The simplest piconet has only two devices.
- The device that initiates the connection is always the master.
- Devices in the same piconet always have access to each other.

Two or more piconets can form a *scatternet* by sharing some common nodes. Any two devices in the same scatternet but in different piconets can access each other. In Figure 2-9, three piconets (nodes enclosed in each ellipse) are connected into a scatternet via shared nodes (the solid gray nodes). Using scatternets, you can extend Bluetooth's coverage. For example, if two piconets' masters are beyond the Bluetooth signal range from each other, they can connect via a third piconet that is physically located between them. In Figure 2-9, nodes B and J can discover and communicate with each other, although they might be outside of each other's range.

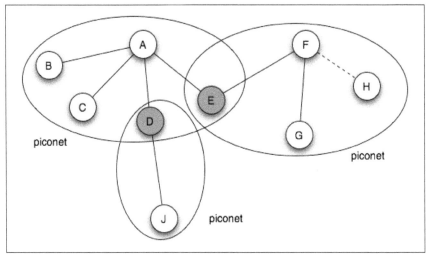

Figure 2-9. Three piconets linked together into a scatternet via shared nodes

> A piconet can have 1 master and up to 7 slaves. A Bluetooth network can contain up to 8 piconets. So, up to 80 Bluetooth devices can connect together in a single network.

The Bluetooth network has its own mechanism to assign dynamic addresses to its devices so that others can find and access them. To find out the unique address of your phone, enter the code *#2820# in standby mode. Alternatively, when you are using a Bluetooth application, you can select which device to connect to using the device's Bluetooth name (refer back to Figure 2-8). This is why you should try to avoid using generic Bluetooth names that might conflict with other names. For example, a name such as "My Phone" or "Nokia3650" is a poor choice.

Bluetooth Services

A Bluetooth device offers services to its peers. To standardize the interaction between Bluetooth devices from different vendors, the Bluetooth specification defines a *service profile* for each service. A service profile specifies a set of data protocols as well as details on how to implement the service on the device. Table 2-1 provides a complete listing of Bluetooth service profiles.

Table 2-1. Bluetooth services

Profile name	Description
Generic Access Profile	Defines how devices discover and connect to each other. It is the basis for all other Bluetooth profiles.
Service Discovery Application Profile	Allows devices to query the services offered by any other device.
Cordless Telephony Profile	Supports Bluetooth-based cordless phones.
Intercom Profile	Allows Bluetooth devices to be used as short-distance voice intercoms.
Serial Port Profile	Defines how two Bluetooth devices can exchange data through serial ports. Since generic binary data can be sent and received via the serial port, this profile is often used for custom Bluetooth applications.
Headset Profile	Allows a phone to be used with a hands-free headset device.
Dial-up Networking Profile	Defines how a device tells a phone to dial a number.
Fax Profile	Defines how a device connects to a fax machine.
LAN Access Profile	Allows one Bluetooth device to act as a network access point for another Bluetooth device.
Generic Object Exchange Profile	Enables applications running on Bluetooth devices to exchange objects.
Object Push Profile	Used together with the Generic Object Exchange Profile, this profile allows one device to push objects to another one, such as a business card.
File Transfer Profile	Defines how two devices exchange files over a Bluetooth network.
Synchronization Profile	Used together with the Generic Object Exchange Profile, this profile supports contacts and calendar item synchronization between devices.

Nokia Bluetooth devices and most computer-based Bluetooth solutions support all the standard service profiles in Table 2-1.

Device Pairing

Due to security concerns, some Bluetooth services are available only to authenticated devices in the network. Bluetooth *pairing* is a process that authenticates two Bluetooth devices to each other. The goal of the pairing process is to have a human being verify that the two devices should trust one another. It works as follows:

1. The first device generates a random PIN number. This number is displayed to you. You can also choose this number yourself and key it into the first device.
2. The first device sends a pairing request to the second device and challenges it for the PIN.
3. The second device asks you for the PIN.
4. You verify that you indeed want to pair those two devices, by typing the PIN into the second device.
5. The second device sends the PIN back to the first device and completes the pairing process.

Figure 2-10 shows how to pair a Nokia mobile phone with a Windows PC. Here the phone is initiating the pairing process. First the phone scans the network for devices, and then it prompts you to enter a pass code. You have to enter the same pass code on the PC to complete the pairing process. Of course, you can also initiate pairing from the Bluetooth software on the PC. I will cover Bluetooth software on Windows, Mac, and Linux later in this hack. After pairing, each device is registered to the other as authenticated, and each can access the other's Bluetooth services.

In Figure 2-10, the first screenshot of the phone shows the Options menu in the Paired Devices screen. On a Nokia Series 60 device, this screen is the second tab in the Connectivity → Bluetooth menu. On a Nokia Series 40 device, you access the list of paired devices via the Settings → Connectivity → Bluetooth → Paired devices menu.

Device and Service Discovery

As I discussed, two Bluetooth devices can connect and use each other's services in several ways.

- They can connect via a master-slave connection.
- Two slaves in the same piconet can connect via the common master.
- Two devices in different piconets but in the same scatternet can connect via shared nodes (refer back to Figure 2-9).

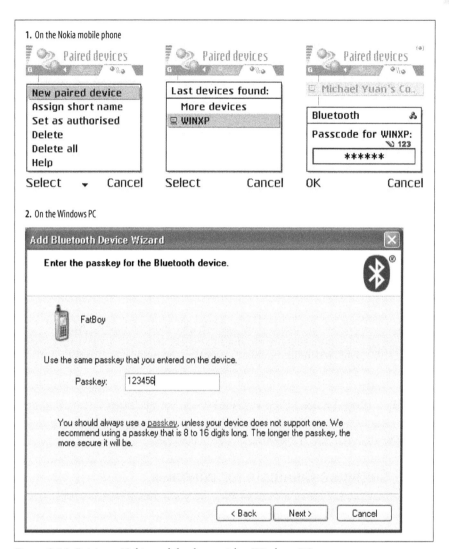

Figure 2-10. Pairing a Nokia mobile phone with a Windows PC

For Bluetooth applications, those connections are transparent. A device uses the Generic Access Profile and Search and Discovery Application Profile to find out its neighboring devices' physical addresses and services. Then it makes a Bluetooth connection to the target address and consumes the services. You can make a device invisible to this discovery process by turning its discovery mode (or visibility) to Hidden (refer back to Figure 2-8). Why would you want to turn off the discovery mode? Well, doing so could reduce your security risk. With the discovery mode on, unknown devices might discover and access your discoverable Bluetooth device in a crowded area (e.g., inside a theater). For example, undiscoverable devices are less vulnerable to bluejack, bluesnarf, bluebug, and backdoor attacks [Hack #22].

If a device were always invisible, no other device in the network would ever find it or use it. In most cases, you make a device discoverable when it first joins a network, and you pair it with some other devices before you turn off the discoverable mode. Paired devices can reach each other without the need for either device to remain discoverable.

Bluetooth Hardware

Many computers, especially Mac laptops, are sold with Bluetooth support built in. But if your computer does not support Bluetooth, you can still add support via third-party Bluetooth kits. A USB Bluetooth dongle is the most popular and easiest to use. Although they are very small and lack a powerful antenna, they can often reach a range of 300 feet (or 100 meters). You also can use a Bluetooth PCMCIA card with your laptop if the laptop has an available PCMCIA slot.

> To use Bluetooth, your Nokia phone must be shipped with Bluetooth support. No third-party kits are available that can add Bluetooth support to older, non-Bluetooth Nokia devices.

In the next several hacks, I'll discuss how you set up Bluetooth pairing and networking between a Nokia smartphone and a computer running Windows [Hack #12], Mac OS X [Hack #13], or Linux [Hack #14].

Configure Bluetooth for Windows
#12
Configure and set up a Windows PC to work with your Nokia smartphone over a Bluetooth connection.

Bluetooth support in the Windows operating system was weak until Windows XP Service Pack 2 (SP2), which significantly improved Bluetooth support. In this hack, I cover Bluetooth for both Windows XP SP2 computers and older Windows computers.

> Windows XP Service Pack 2 is a free upgrade for all Windows XP computers that includes greatly enhanced security and better WiFi and Bluetooth support. I highly recommend that you download and install it.

Once you have the Bluetooth connection set up, you can follow the instructions in "Use Bluetooth to Replace Cables" **[Hack #11]** to pair the phone with the PC.

Windows Versions Earlier Than Windows XP SP2

In pre-SP2 versions of Windows, including Windows 2000, you have to install the vendor-supplied software for the Bluetooth adapter. The software typically provides both a device driver and Bluetooth service management tools for this adapter. Many different kinds of Bluetooth adapters are available, and I cannot cover them all. So, in this section, I'll use the WIDCOMM software, which is used with many USB Bluetooth dongles, as an example.

Most Bluetooth adapters on the market use the WIDCOMM Bluetooth software for Windows PCs. Hence, all these adapters have a similar look and feel from the user's perspective. However, different hardware manufacturers use their own customized version of the WIDCOMM software, so you generally cannot use one manufacturer's driver with another manufacturer's adapter. You can find the vendor-specific Bluetooth driver in the CD that comes with your device. Or, you can often download the latest driver for free from the vendor's web site. You can get more information about the WIDCOMM drivers from *http://www.broadcom.com/*.

Once you install the WIDCOMM software, it adds a new My Bluetooth Places shortcut to each user's desktop and a Bluetooth icon to the system tray. My Bluetooth Places opens like a folder in Windows Explorer. The content in the folder shows the devices in the network and the services available. The available Bluetooth operations, such as searching for new devices and starting/stopping local services, are listed in the Bluetooth Tasks sidebar. The tasks are context sensitive. For example, a different set of Bluetooth tasks is available when you are browsing connected devices than when you are browsing the services on the local computer (see Figure 2-11).

You can start the system Bluetooth configuration utility from the system tray icon or via the "View or modify configuration" task in the My Bluetooth Places browser (see Figure 2-12). Through the configuration utility, you can edit your Bluetooth settings, search for devices in the network, and pair with devices. You can also use the vendor tools to use services on paired mobile phones, such as dialing numbers, exchanging files and business cards, and so forth.

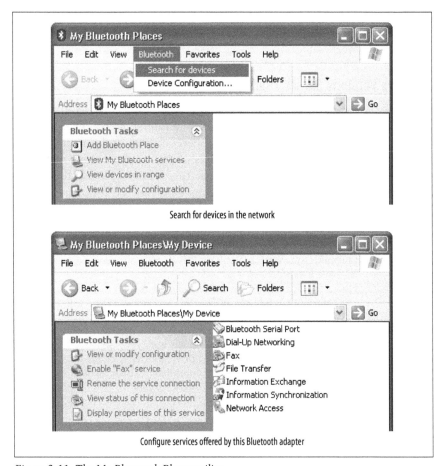

Figure 2-11. The My Bluetooth Places utility

Windows XP with SP2

Windows XP SP2 includes Bluetooth device drivers for several popular Bluetooth adapters, including cards and USB dongles. Once you plug in a supported Bluetooth adapter, the computer will recognize it and install it (if it doesn't, use the drivers supplied by your vendor). After installation, you will see new Bluetooth icons in the Network and Internet Connections section of the Control Panel, as well as in the system tray. You can double-click either of these two icons to bring up the Bluetooth configuration utility. From here, you can manage the computer's Bluetooth settings, such as discovery mode and permission settings (see Figure 2-13). You also can use it to discover and pair with devices in your network. In Figure 2-14, the left window shows the paired devices. You can click the Add button to pair more

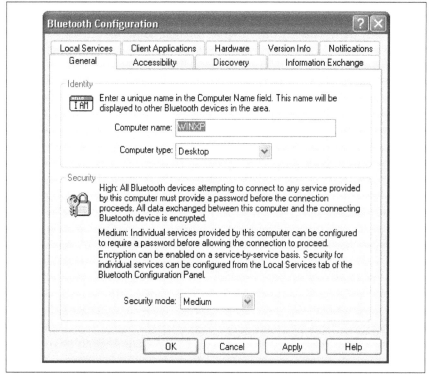

Figure 2-12. The configuration tool for WIDCOMM Bluetooth adapters

devices. Once a device is paired, you can inspect the serial ports through which it is connected to the computer by clicking the Properties button (this brings up the righthand window in Figure 2-14). The serial port information is important for a variety of Bluetooth applications.

> Check the "Show the Bluetooth icon in the notification area" box in the Bluetooth Options (see the left window in Figure 2-13) to keep the Bluetooth icon in the system tray. If you already have the WIDCOMM driver installed, you will have two Bluetooth icons in the system tray now: the Windows XP SP2 icon, and the WIDCOMM icon. The Windows XP SP2 native icon is smaller in size.

If Windows XP SP2 cannot recognize your Bluetooth adapter, you need to install the vendor-specific software driver (follow the instructions in the "Windows Versions Earlier Than Windows XP SP2" section of this hack). The next time the Bluetooth adapter is plugged in the computer will recognize it and find the appropriate driver. Windows XP SP2's default Bluetooth configuration utilities override all vendor-provided management tools.

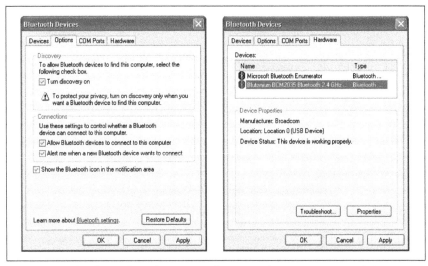

Figure 2-13. The Bluetooth configuration utility in Windows XP Service Pack 2

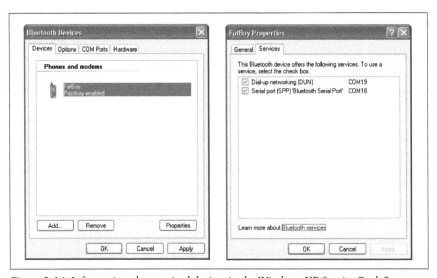

Figure 2-14. Information about paired devices in the Windows XP Service Pack 2
Bluetooth management tool

H A C K
#13 Configure Bluetooth for Mac OS X

Configure and set up a Mac OS X computer to work with your Nokia
smartphone over a Bluetooth connection.

Mac OS X provides excellent support for Bluetooth. Most new Mac laptops
are shipped with Bluetooth preinstalled and preconfigured. To check

whether yours has built-in Bluetooth, you can open System Preferences and check whether the Bluetooth icon exists. If your Mac does not have built-in Bluetooth, you can simply insert a compatible USB Bluetooth dongle. Mac OS X recognizes most Bluetooth dongles and automatically adds the Bluetooth icon to System Preferences. If it does not, it is unlikely that you will be able to obtain drivers from the vendor. If you click the Bluetooth icon, the Bluetooth preferences pane opens. In Figure 2-15, the top window shows the Bluetooth properties of this Mac and the lower window shows the paired devices. Click the Set Up New Device button to search for and pair with your phone **[Hack #11]**.

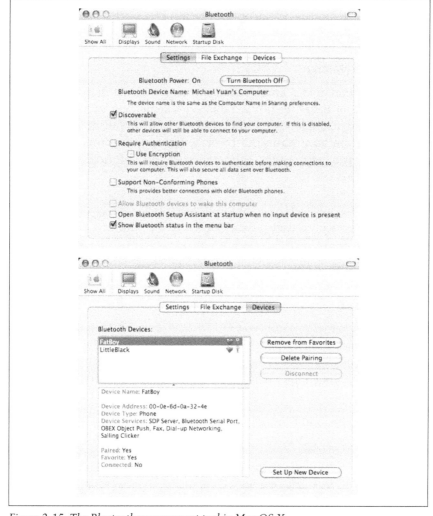

Figure 2-15. The Bluetooth management tool in Mac OS X

Configure Bluetooth for Linux

#14   Configure and set up a Linux computer to work with your Nokia smartphone over a Bluetooth connection.

Bluetooth support on Linux is a complex issue. As with many emerging technologies, competing implementations of Linux Bluetooth support exist. The two main implementations are Affix and BlueZ. Affix was developed by Nokia and is now hosted as an open source project at SourceForge (*http://affix.sourceforge.net/*). BlueZ is also available (*http://www.bluez.org/*) and is the official Bluetooth stack of the Linux kernel.

Although Affix is a mature and functional project, BlueZ receives more testing and has been more widely adopted. For this reason, this hack focuses on the uses of the BlueZ Linux Bluetooth stack and libraries.

Kernel Requirements

Bluetooth support under Linux requires a recent kernel. If your kernel is Version 2.4.22 or better, or if it is a 2.6 series kernel, you're all set. Otherwise, you must upgrade your kernel. Alternatively, if you do not want to upgrade, and you have kernel 2.4.18 or better compiled from source, you can apply the patches from the "kernel patches" area of the BlueZ web site (*http://www.bluez.org/*).

> Patching and recompiling the kernel for Bluetooth support is not a trivial matter. For a complete set of instructions, please refer to Chapter 7 of *Linux Unwired* (O'Reilly, 2004). I recommend that you use a recent kernel rather than a patch, if at all possible.

Installing BlueZ

In addition to kernel support, you must install a set of utility programs to help you manage your Bluetooth devices. Table 2-2 shows the name of the packages and their purpose. You can either install the versions of these tools that come with your Linux distribution or compile and install them from source.

Table 2-2. Utility programs for managing Bluetooth devices

Package	Purpose
bluez-libs	The application library that all other Bluetooth tools require to function
bluez-utils	Main utilities that enable you to initialize and control Bluetooth devices
bluez-sdp	Service discovery protocol tools that enable the advertisement and discovery of Bluetooth services

Table 2-2. Utility programs for managing Bluetooth devices (continued)

Package	Purpose
bluez-pan	Tools that enable personal area networking using Bluetooth
bluez-hcidump	A debugging tool that permits the monitoring of Bluetooth packets
bluez-bluefw	The firmware for Broadcom chipset-based Bluetooth devices

If you are compiling the tools from source code, compile and install in the order shown in Table 2-2 to avoid dependency problems.

You can obtain precompiled versions of the utilities for Red Hat Linux as RPMs, for Debian stable as *.deb* packages (the latest BlueZ utilities are an integral part of Debian unstable), and as packages suitable for the Sharp Zaurus Linux PDA. You can download them, along with the source code packages, from the BlueZ download page at *http://www.bluez.org/*.

To determine whether your Bluetooth system is working, you need to install only the bluez-libs and bluez-utils packages, and also bluez-bluefw if your dongle contains a Broadcom chip (you can determine this from Marcel Holtmann's Bluetooth hardware page at *http://www.holtmann.org/linux/ bluetooth/devices.html*). Install the rest when you have verified that everything is working properly.

Some Basic Command-Line Operations

The bluez-utils package contains the tools you need to configure and test your Bluetooth setup. Once you've installed the package, run the init script (*/etc/init.d/bluez-utils start* on Debian, */etc/init.d/bluetooth start* on Red Hat) to start the Bluetooth subsystem. These scripts normally run on boot, so they might have been started already if you installed from RPMs or Debian packages.

The hcid daemon should now be running. This program controls the initialization of Bluetooth devices on the system and handles the pairing process with other devices. I discuss hcid configuration later in this chapter.

The prefix *hci* derives from the name of the interface between the computer and the Bluetooth device, or the Host Controller Interface.

In this section, I'll show you some of the most basic Bluetooth operations using the BlueZ command-line tools to verify that the system does work. For more operations, please refer to Chapter 7 of *Linux Unwired* (O'Reilly, 2004).

Examining local devices. The `hciconfig` tool allows the configuration of the characteristics of your Bluetooth adapter. If you are familiar with the configuration of network interfaces, you will find it parallel in operation to `ifconfig`. Use -a to display extended information about each Bluetooth device attached to the computer:

```
# hciconfig -a
hci0:   Type: USB
        BD Address: 00:80:98:24:15:6D ACL MTU: 128:8  SCO MTU: 64:8
        UP RUNNING PSCAN ISCAN
        RX bytes:4923 acl:129 sco:0 events:168 errors:0
        TX bytes:2326 acl:87 sco:0 commands:40 errors:0
        Features: 0xff 0xff 0x05 0x00
        Packet type: DM1 DM3 DM5 DH1 DH3 DH5 HV1 HV2 HV3
        Link policy: HOLD SNIFF PARK
        Link mode: SLAVE ACCEPT
        Name: 'saag-0'
        Class: 0x100100
        Service Classes: Object Transfer
        Device Class: Computer, Uncategorized
        HCI Ver: 1.1 (0x1) HCI Rev: 0x73 LMP Ver: 1.1 (0x1) LMP Subver: 0x73
        Manufacturer: Cambridge Silicon Radio (10)
```

Scanning for remote devices. The acid test, of course, is to see if your computer can detect other Bluetooth devices. You can use the `hcitool` tool to do this. Switch on your other Bluetooth device, and ensure it is in "discoverable" mode. Issue the command hcitool `scan` and wait. You don't need to be root to run this command.

```
$ hcitool scan
Scanning ...
        00:0A:D9:15:CB:B4       ED P800
        00:40:05:D0:DD:69       saag-1
```

The previous listing shows a typical output of a scan. In this case the author's cell phone, ED P800, and second Bluetooth adapter, saag-1, are shown as discoverable.

Pairing. Many devices require that pairing be performed before a Bluetooth connection is established. The computer or the remote device can initiate pairing.

If the computer initiates pairing—usually by making an outgoing connection—the *pin_helper* program (usually *bluepin*) will present a graphical dialog box to the user requesting that he input a PIN, which should match the code set on the remote device. If the remote device initiates pairing, the remote device is required to provide a PIN to match the contents of the */etc/bluetooth/pin* file.

In some distributions of bluez-utils, the PIN code is set to the alphabetical string *BlueZ*. This is troublesome, because many Bluetooth devices, including most cell phones, are capable of delivering only numeric PINs. It is therefore recommended that you alter the contents of */etc/bluetooth/pin* to reflect a numeric code.

If pairing is successful, the hcid daemon will store the resulting link key, used to authenticate all future connections between the two devices concerned, in the */etc/bluetooth/link_key* database file.

Pinging a remote device. The ping command is an incredibly useful tool for discovering whether remote computers are reachable over a TCP/IP network. BlueZ has an analog to ping, called l2ping. Its name refers to the fact that it attempts to create a connection to the device using the logical link control and adaptation protocol (L2CAP), the lowest-level link-based protocol in Bluetooth.

In other words, before despairing because you cannot connect to a device, check it with l2ping. There might be a fault with software higher up the chain; l2ping enables you to determine whether a basic connection can be established with a remote device. Here's an example of l2ping in action (you need to run l2ping as root):

```
# l2ping 00:0A:D9:15:CB:B4
Ping: 00:0A:D9:15:CB:B4 from 00:80:98:24:15:6D (data size 20) ...
0 bytes from 00:0A:D9:15:CB:B4 id 200 time 54.85ms
0 bytes from 00:0A:D9:15:CB:B4 id 201 time 49.35ms
0 bytes from 00:0A:D9:15:CB:B4 id 202 time 34.35ms
0 bytes from 00:0A:D9:15:CB:B4 id 203 time 28.33ms
4 sent, 4 received, 0% loss
```

GUI Bluetooth Applications

Linux has several popular graphical user interface (GUI) systems, the most well known being KDE and GNOME. Both of these projects have tools that provide an easy-to-use interface to your system's Bluetooth devices. At the time of this writing, neither project is an official part of the KDE or GNOME desktop, but both will be integrated in the future. This section presents a brief survey of the tools available, and where to get them.

Before you can use the GUI frontends for Bluetooth on Linux, you need to install and configure BlueZ.

KDE. The KDE Bluetooth Framework's home page is at *http://kde-bluetooth. sourceforge.net/*. Its features include:

- A control center plug-in to configure Bluetooth devices
- An OBEX server application
- An OBEX sending client
- Graphical exploration of remote devices
- Cell phone hands-free implementation using your computer's microphone and speakers
- Proximity-based screen locking

You can download the KDE Bluetooth Framework from the project's web page. Figure 2-16 shows KDE's Bluetooth applications in action.

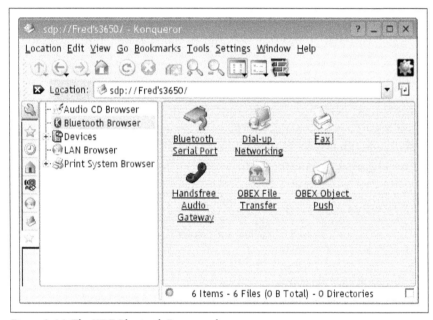

Figure 2-16. The KDE Bluetooth Framework

GNOME. The GNOME Bluetooth subsystem's home page is available at *http:// usefulinc.com/software/gnome-bluetooth*. Download it from the project's home page. RPM and Debian packages are also available. Features of the GNOME Bluetooth subsystem include:

- An OBEX server application
- An OBEX sending client
- A phone manager application allowing sending and receiving of Short Message Service (SMS) messages

- Graphical exploration of remote devices
- Programming libraries for creating Bluetooth-aware applications in C, Python, or C#

Figure 2-17 shows GNOME's Bluetooth features in action.

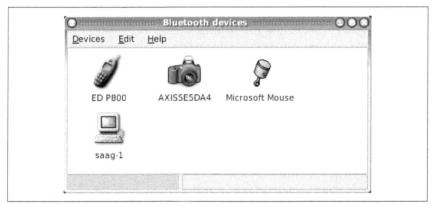

Figure 2-17. The GNOME Bluetooth subsystem

—Brian Jepson and Edd Dumbill

Use the Nokia PC Suite

The Nokia PC Suite allows you to use a PC to control, manage, and extend your Nokia smartphone. It greatly enhances your device experience.

Today's smartphones have computing power similar to PCs. However, due to their small screens and limited keyboards, they are not about to replace PCs. In fact, the phone and PC usage models nicely complement each other. You can do the heavyweight computing (e.g., writing and compiling code, creating gigantic spreadsheets, or designing a company newsletter) on a PC and have mobile data access on your smartphone anywhere you go. The phone is a smart extension to the PC. The Nokia PC Suite lets your Windows PC work with your Nokia phone. Many hacks in the rest of the book use the Nokia PC Suite.

Download and Install the PC Suite

The Nokia PC Suite software is freely available from the Nokia web site, at *http://www.nokia.com/pcsuite.* You will be asked to choose your phone model, and the Windows operating system version on which you plan to run the PC Suite. Depending on your choices, you will be presented with several possibilities. All of the version numbers in the following list are current as of April

2005. For a complete compatibility table of devices and PC Suite versions, please visit *http://www.nokia.com/nokia/0,,72030,00.html#model*.

- Most Nokia phones work with both the Nokia PC Suite v6.5 (for Windows 2000 and XP) and v5.8 (for Windows 98 and ME). The v5.8 software is provided only to support older Windows computers, and is no longer under active development. It is not as feature-rich as v6.5. Some phone models, including the popular Nokia 7610 (Series 60 **[Hack #2]**), work only with the Nokia PC Suite v6.5. You cannot install a Nokia PC Suite for those devices on a Windows 98/ME computer.

- Some Nokia devices, including the popular Nokia 3650 and 6600 (Series 60), require special versions of the Nokia PC Suite (e.g., the Nokia PC Suite for 6600). Those special versions work on Windows 98, ME, 2000, and XP. However, some of them have known problems with Windows XP SP2's native Bluetooth drivers.

- Some older phone models (black-and-white screen models) work only with the Nokia PC Suite v4.8.x, which runs on Windows 2000, XP, 98, and ME computers. Those devices and their PC Suites have limited functionalities.

In this book, I will focus on the Nokia PC Suite v6.5. If your phone or operating system requires you to use another version, please refer to the corresponding user manual and you can probably find features similar to those in v6.5.

Once you download the Nokia PC Suite installer, you can launch it and follow the on-screen instructions to install the program. If you want to upgrade your Nokia PC Suite to a more recent version, you have to uninstall the old version first via the Add/Remove Programs control panel and then install the new version.

 Many problems with the upgraded PC Suite are caused by Windows registry problems that occurred during the uninstallation of the old version. The Nokia Registry Cleaner utility (freely available from the same page where you downloaded the Nokia PC Suite) removes unnecessary debris from the Windows registry.

A Tour of Key Features

The Nokia PC Suite consists of a collection of programs. Each icon in the PC Suite's main window corresponds to a separate program (see Figure 2-18). Table 2-3 lists the component programs in the Nokia PC Suite v6.4.

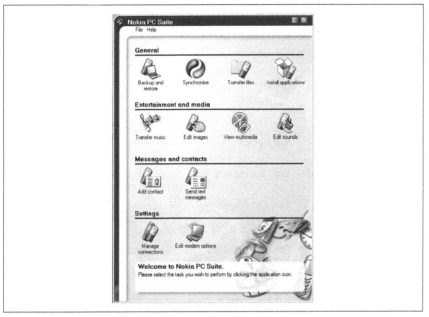

Figure 2-18. The main window of the Nokia PC Suite v6.4

Table 2-3. Programs in the Nokia PC Suite v6.4

Program	Icon name	Description
Nokia Content Copier	Backup and restore	Backs up and restores data from and to the mobile device. It can also be used to copy content from one device to another.
Nokia PC Sync	Synchronise	Synchronizes personal information items (contacts, calendar events, and to-do items) between the device and the Outlook or Lotus programs on the PC.
Nokia Phone Browser	Transfer files	Allows you to browse the content on the phone using Windows Explorer on the PC.
Nokia Application Installer	Install applications	Installs Java and Symbian applications to supported devices.
Nokia Audio Manager	Transfer music	Creates and edits audio clips in formats supported by Nokia devices. It also allows you to transfer audio clips between the device and the PC.
Nokia Image Converter	Edit images	Converts images to formats supported by Nokia devices. It also allows you to transfer images to the device to be used as wallpaper, MMS messages, etc.

Table 2-3. Programs in the Nokia PC Suite v6.4 (continued)

Program	Icon name	Description
Nokia Multimedia Player	View multimedia	Plays video clips taken by the phone's camera on the PC.
Nokia Sound Converter	Edit sounds	Converts generic Musical Instrument Digital Interface (MIDI) ring-tone files to special MIDI ring-tone formats supported by Nokia devices.
Nokia Contacts Editor	Add contact	Enters contacts on the PC and sends them over to the phone's contacts list.
Nokia Text Message Editor	Send text messages	Sends and receives text messages on the PC via the phone.
Nokia Connection Manager	Manage connections	Manages the devices connected to this PC via Bluetooth, Infrared, USB cable, and serial cable.
Nokia Modem Options	Edit modem options	Enables the PC to use the mobile phone as a modem.

Connect the PC Suite to a Device

The Nokia PC Suite can access devices connected to the PC via Bluetooth, Infrared data port, serial cable, and USB cable. If you prefer to use a cable-based connection, you have to purchase the cable separately (cables are available via many online stores). Refer to your phone's specification in the manual to determine the type of cable it supports before purchasing.

To connect a device to the PC Suite, you can click the Manage Connections icon in the main window to bring up the Connection Manager application (see Figure 2-19). Then you can select the connection method for the device and click the Configure button to configure the connection. For example, if you click the Configure button for Bluetooth connection, the PC Suite will walk you through a wizard to search, pair, and connect to nearby Nokia phones.

PC Suite Limitations

The Nokia PC Suite has many useful features that are used by several other hacks in this book. However, sometimes you might not be able or willing to use the PC Suite to connect to the device:

- The Nokia PC Suite is available only on Windows PCs. It does not work on Mac or Linux computers.
- While the Nokia PC Suite v6.5 is the most recent version of the software, it runs only on Windows 2000 and XP computers. If you use older Windows PCs, you will have to settle for older versions of the PC Suite that have fewer features and are no longer under development.

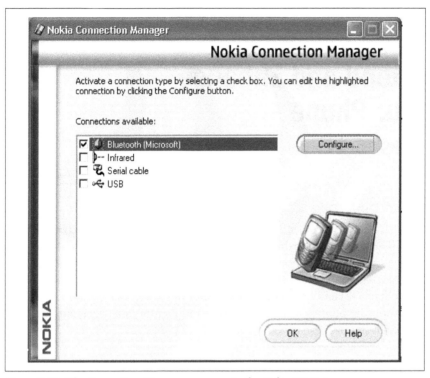

Figure 2-19. The Connection Manager program in the Nokia PC Suite

- Not all devices work with the Nokia PC Suite v6.5. Several popular devices (e.g., the Nokia 3650 and 6600) require their own special versions of the PC Suite. Yet, you cannot run two Nokia PC Suites on the same computer. This is inconvenient for users with multiple Nokia devices.

- The Nokia PC Suite is sometimes buggy. That is especially the case for the Windows 98–based PC Suite, and for PC Suites that target specific devices. Apparently, those versions of the Nokia PC Suites receive less development and Q/A effort than the main v6.5 version. For example, some versions of the Nokia PC Suite do not work with the native Bluetooth stack on Windows XP SP2 **[Hack #12]**.

- The Nokia PC Suite is a 20MB application that consumes a significant amount of system resources on the PC.

- Mobile users sometimes need to use public computers with the phone. It is impractical to install the Nokia PC Suite on public computers.

In the rest of this book, when I describe a hack using the Nokia PC Suite, I'll discuss any alternative ways in which you can complete the same task.

CHAPTER THREE

Extend and Enhance
Your Phone
Hacks 16–21

A key characteristic of smartphones is that they can run third-party applications. Those applications add new features to the phone, essentially allowing you to extend the phone beyond its original capabilities. In this chapter, I'll cover techniques to install, run, and manage third-party applications on Nokia Series 40 and Series 60 smartphones. Issues related to running applications on the phone, such as filesystem and memory management, are also covered in the hacks in this chapter.

HACK
#16 Run Java Applications

Your phone has a built-in Java environment that lets you run all sorts of cool applications and games. Installing and managing these applications is easy, once you know how.

All Nokia Series 40 and Series 60 phones support Java applications (a.k.a. MIDlets) that you can add on to your phone after you get it. With these applications, you can customize, enhance, and extend your phone to suit your preferences.

> A MIDlet is a Java application that conforms to the Mobile Information Device Profile (MIDP) standard. The MIDP specification defines the Java Virtual Machine and API available on most Java-compatible smartphones, including all Nokia smartphones.

Java technology is most widely used to develop mobile games and entertainment applications for Nokia phones. In fact, your Nokia phone probably was preinstalled with some popular Java games (e.g., Golf and Beach Rally) and utilities (e.g., Converter and World Clock), courtesy of Nokia and your wireless operator. Here is a list of web sites where you can purchase or download MIDlets:

- Handango is the world's largest online mobile software store. It runs Nokia and Motorola's online software stores. Visit Handango at *http://www.handango.com/*.

- The mpowerplayer web site (*http://www.mpowerplayer.com/*) provides "previews" of mobile Java games right on your desktop computer. You can download any game and play it on your computer. You need only to pay for the game once you decide to install it onto your phone. It is extremely cool. Check it out!

- The Midlet Review (*http://www.midlet-review.com*) publishes reports and reviews of new MIDlets, especially games.

A Java application consists of two files. The Java archive (JAR) file contains the application's executable code. The Java application descriptor (JAD) file is a text file that contains attributes about the application and the location of the JAR file. During the installation process, the phone first grabs the JAD file, parses its content, and checks whether the device has the required software and memory space to install the application. If everything looks OK, the device follows the URL in the JAD file to locate and download the JAR file. Then the device compares the JAR file with attributes in the JAD file (e.g., application name, vendor, size, digital signature, etc.). If everything matches, the application is installed and becomes available to the user. This hack covers how to install and use Java applications on your phone.

Install over the Air

The easiest way to install a Java application is to download it via the Internet. This is called Over The Air (OTA) installation. To do this, you need to use your phone's native WAP browser (the Services application). You can point the browser to a web site that lists mobile Java applications, and then click the download link for the application you want to install. The browser downloads the JAD file and starts the Java Application Management Software (AMS) to process it. The AMS prompts you to confirm a series of choices, automatically downloads the JAR file, and takes care of the rest of the installation process (see Figure 3-1). Notice that you cannot use any of the non-Nokia-native web browsers discussed in "Browse the Web" [Hack #50], since they do not know how to invoke the AMS. For the OTA installation to work, the web server that offers the download must map the *.jad* file suffix to the *text/vnd.sun.j2me.app-descriptor* Multipurpose Internet Mail Extensions (MIME) type and the *.jar* file suffix to the *application/java-archive* MIME type [Hack #53].

You can purchase or download Java applications from many web sites. Your WAP browser's bookmark section probably already has links to Nokia and

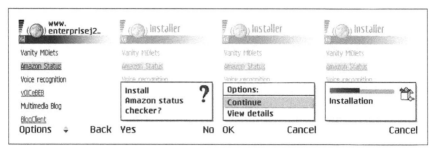

Figure 3-1. Installing Java applications over the air

your wireless operator's Java application catalog web pages. Or, you can try one of the large independent sites for Java applications, such as *http://www. handango.com/*, *http://www.midlet.org/*, or *http://www.java.net/*. You can also easily set up your own web site to provision Java applications **[Hack #53]**.

Install from Your Computer

In some cases, you might want to install a Java application directly from your PC to the phone. You might have gotten the JAD and JAR files via email, downloaded them manually from a web site, or written them yourself. Or, maybe your phone does not have General Packet Radio Service (GPRS) or WAP services **[Hack #10]** to download the applications (or you just don't want to use up your bandwidth allotment with these services). In those cases, you can use the Nokia PC Suite to install a pair of matching JAD and JAR files directly from your Windows PC to the phone (see Figure 3-2). The Nokia Application Installer program in the PC Suite allows you to choose a JAR file from the local hard drive. Once you click the green arrow to install the JAR file to the phone, the PC Suite searches the directory the JAR file resides in for a matching JAD file of the same name. If such a JAD file is found, the JAD and JAR pair will be installed automatically. If the JAD file is not found, you can install the JAR file by itself (the JAR-only installation).

If you use a Mac computer or simply cannot get the Nokia PC Suite working on your computer, you can send the JAR file over to the phone via a local connection, such as Bluetooth or Infrared **[Hack #33]**. The file shows up in your phone's messaging Inbox. When you open the message, the AMS is invoked to install the application (see Figure 3-3 for an example of installation over Bluetooth). This method is one way to perform a JAR-only installation.

Figure 3-2. Installing a JAR file and its corresponding JAD file via the Nokia PC Suite

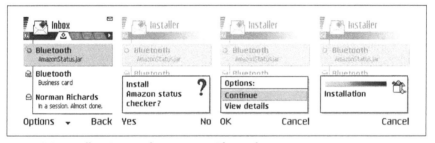

Figure 3-3. Installing Java applications over Bluetooth

The JAR-only installation method is not standard and does not work on all Nokia devices. You have to try it with your device to see if it works. Even if it does, the JAR-only installation method might have some potentially serious shortcomings. Since the JAD file is not passed to the device, some crucial application attributes might get lost. For example, the device will not be able to authenticate the application or grant it security permissions without the JAD file. If the application relies on custom-defined JAD attributes to function at runtime, it might fail or behave incorrectly.

Run Java Applications

On Series 60 devices released after 2003 (e.g., the Nokia 6600 and later), the AMS installs MIDlets as top-level icons in the Main menu. You can move them to other folders as you wish. On older Series 60 devices (e.g., the

Nokia 3650), Java MIDlets are accessible only via the Games or Apps. icon in the Main menu (described in the next section). For Series 40 devices, you can find installed Java applications under the Applications → Collection or Applications → Games menu. To start a Java application, simply open it (select it and press the center button on the navigation pad, or press the Options soft key and select Open from the menu that appears).

Manage Java Applications

On newer Series 60 devices (e.g., the Nokia 6600 and later), the native Application Manager (available from the Main menu) lists all the user-installed applications. You can navigate to any application and select the Options soft key to bring up the management menu for it. Through this menu, you can delete the application, access its update history, and automatically update the application to a newer version if its download web address is known. Via the Settings menu, you can alter how the application asks for user permissions before performing sensitive actions. For example, you can configure the application to prompt you for confirmation every time it accesses the Internet or sends messages (see Figure 3-4).

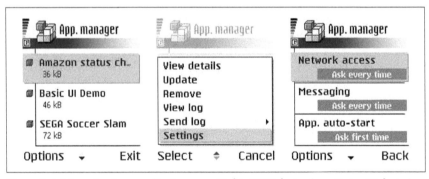

Figure 3-4. Managing Java applications on a Nokia 6600 device (new Series 60 device)

As noted earlier, on older Series 60 devices (e.g., the Nokia 3650) you can access the MIDlet settings from the Options menu when you highlight the MIDlet in the Games or Apps. menu (see Figure 3-5). Note that these older devices offer fewer configurable options than newer Series 60 devices such as the Nokia 6600, since they conform to an earlier version of the J2ME/MIDP specification.

On a Series 40 device, the application management functions are accessible from the Options menu associated with each application under the Applications → Collection or Application → Games menu. Instead of the Settings menu, you'll use the "App. access" menu on a Series 40 device (see Figure 3-6).

Figure 3-5. Accessing and configuring Java MIDlets in a Nokia 3650 device (old Series 60 device)

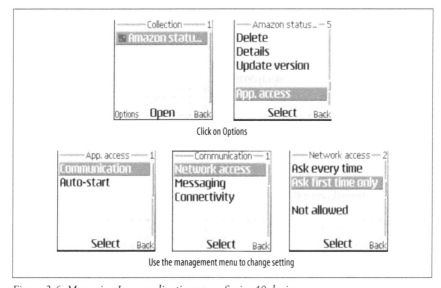

Figure 3-6. Managing Java applications on a Series 40 device

For serious phone hackers, Java provides a way (the *only* way for Series 40 devices) to enhance your device via a full-featured programming language. In this hack, I cover how to install and manage Java applications on your phone. The techniques discussed in this hack are used in many other hacks throughout this book.

Write Your Own Java Applications

If you know Java, it is easy to write your own Nokia smartphone applications using J2ME. To do that, you need the Nokia Developer's Suite for J2ME, which is a free download from the Forum Nokia web site (*http://www.forum.nokia.com/*). The Nokia Developer's Suite for J2ME runs on

Windows and Linux computers and provides the following features from a simple graphical user interface (GUI):

- A Java source code editor
- A drag-and-drop user interface (UI) builder to generate UI classes for the phone application from a visual designer
- Tools to compile Java source code and build the JAD and JAR files for distribution
- Device emulators to test your applications on the computer before you deploy them to the physical device

Since Nokia makes many phones, there are several device emulators to choose from. The Nokia PC Suite for J2ME comes bundled with a Nokia 6230 emulator and a generic Nokia Series 60 phone emulator. You can download and install more emulators for free from the Forum Nokia web site. The Nokia Developer's Suite for J2ME manages all installed emulators.

If you are an experienced Java developer and prefer to work with Integrated Development Environment (IDE) tools, you can install the Nokia Developer's Suite for J2ME inside a popular IDE such as Eclipse. This way, the J2ME development features appear inside the IDE's menu instead of in a standalone GUI.

For more information about Java application development on Nokia Series 40 and Series 60 smartphones, please refer to the book *Developing Scalable Nokia Series 40 Applications* (Addison Wesley, 2004).

> While the Nokia Developer's Suite for J2ME does not run on Mac computers, Mac OS X users can use an open source J2ME development toolkit, including a generic device emulator, developed by Michael Powers. Check it out here: *http:// www.mpowerplayer.com/for_developers.php*.

HACK #17 Run Symbian Applications

Enhance your Series 60 phone with Symbian OS applications.

Nokia Series 60 devices are based on the Symbian operating system (OS). They can run native Symbian applications in addition to Java MIDlets **[Hack #16]**. Compared with Java, native Symbian applications integrate much better into the underlying phone system. In fact, most of the phone's built-in applications, such as the call dialer, the messaging client, and the browser, are Symbian applications. However, native Symbian applications are also more difficult to develop than Java, and hence, fewer of them are available. Most third-party native Symbian applications are commercial applications.

Here is a list of places where you can purchase or download Symbian applications for your Nokia phone:

- Handango is the world's largest online mobile software store. It powers Nokia and Motorola's online software stores. You can visit Handango at *http://www.handango.com/*.

- My-Symbian.com (*http://my-symbian.com/*) publishes news about the latest Symbian software. It sells Nokia Series 60 and Series 80 Symbian applications via its online store.

- SymbianWare (*http://www.symbianware.com/*) offers Symbian applications for Nokia devices. Some applications in its catalog are available for free download.

- The SYMBOS software store (*http://www.symbos.com/*) sells Symbian and Java software for Nokia's Symbian OS phones.

A Symbian application is typically distributed in a single installation file with a *.sis* filename suffix. All the executable code, resource files, and metadata are bundled in the *.sis* file. While it is possible to install the Symbian *.sis* file directly over the air from your mobile browser, most Symbian application download sites do not have the correct MIME type (the *application/vnd.symbian.install* type) associated with the *.sis* files. If you point your mobile browser to a *.sis* file on those sites, the browser just treats it as if it is a text file and displays it in the phone's Note editor program (see Figure 3-7).

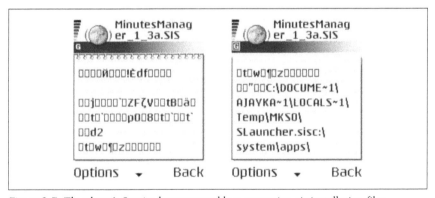

Figure 3-7. The phone's Service browser unable to recognize .sis installation files

Install a Symbian Application

The best way to install a Symbian application is to first download the *.sis* installation file to your computer and then pass it from the computer to the device. Unlike Java applications, which are typically distributed by the wireless operator, Symbian applications are available via independent software stores or the developer's own web site. You can navigate to the sites via your computer browser and select the software product. Once you agree to the end-user license (or purchase the software), you will be presented with a direct link to the *.sis* installation file. Click the link and your browser starts to download the file. In most cases, the browser will not recognize the type or the content of the *.sis* file. So, you are likely to see random characters in the main browser window. But don't worry; just use the File → Save As menu to save the file into a local file with the *.sis* filename suffix once downloading is finished.

At this point you can run the Nokia PC Suite to install the *.sis* file on the device via a local connection. However, if you have a Bluetooth or Infrared connection set up between the computer and the device, it is probably easier to just beam **[Hack #33]** the *.sis* file to the device. The file shows up as a new message in the device's system Inbox. Once you open that message, the device installs the Symbian application. You will be asked to confirm a series of choices to make sure you understand what's going on before the application is actually installed (see Figure 3-8).

Figure 3-8. Installing Symbian applications via a Bluetooth connection

The installed Symbian application shows up in the device's Main menu. You can start it by opening it. You can also move it to other folders via the Options menu.

In addition, you can purchase memory cards that are preloaded with Symbian applications. In this case, no installation is necessary. The applications become available on the Main menu (see Figure 3-9) after you power down your phone, insert the memory card, and restart the phone.

Register a Symbian Application

Most Symbian applications are commercial software. To reduce piracy, they typically require that you activate the software after it is installed. If you do not activate the application within the specified trial period, it will stop

Figure 3-9. Symbian applications preinstalled on a MultiMediaCard (MMC) card

working. The activation process ties the software purchase to a specific device. It works as follows:

1. Each Global System for Mobile communications (GSM) device is uniquely identified by a 15-digit International Mobile Equipment Identity (IMEI) code. You can get your device's IMEI code by typing *#06# on the keypad **[Hack #5]**.

2. Write down the IMEI code of the device you installed the software on and submit it to the software store or the developer. The option to submit the code will be made available to you when you download or purchase the *.sis* file.

3. The vendor or developer then sends you back an activation code (usually within minutes, but sometimes within a few days). You bring up the Activate menu of the application and type in the code to complete the activation process.

The Symbian application uses a secret algorithm (probably a secure hash algorithm with a secret key) to calculate the expected activation code based on the IMEI code. The software store and the developer know how to calculate the activation code from the IMEI code as well. In the meantime, as the user, you do not know the secret and hence cannot calculate the activation code yourself. So, if you supply the correct activation code to the application, it "knows" that you have gone through the official activation process. If you need to use the application on a different device beyond the grace period, you might have to repurchase the product and get a separate activation code. However, you should contact the vendor and explain that you are switching phones. They might offer you a new activation code for little or no cost.

The activation process enables Symbian software developers to offer trial versions of their software. For most commercial applications, you can download and install the *.sis* file for free and try it out during the trial period. You need to purchase the software only after you verify that it works with your device and are satisfied with its performance. The trial period could last a number of days or a number of restarts after you first install the application. In fact, Nokia Series 60 devices often come with trial versions of Symbian applications preinstalled on their add-on memory cards. They typically allow you to use them 5–10 times before activation.

Manage Symbian Applications

Just as you can manage Java MIDlets, you can manage Symbian applications via the Manager application, which is accessible via the Main menu (on older Series 60 devices, you'll find it in the *Tools* folder off of the Main menu). Once you start the Manager application, you will see a list of Java MIDlets and Symbian applications installed on the device. You can move the scroll key to highlight any of them and use the Options menu to view their memory details and digital certificates. You can also remove any installed applications using this menu.

Write Your Own Symbian Applications

If you are familiar with the C++ language, it is fairly quick to learn how to write Symbian C++ applications for Nokia Series 60 devices. Compared with regular C++, Symbian C++ requires a complex build structure with a lot of resource files and configuration files. A good IDE tool hides those complexities, and hence, greatly reduces the Symbian C++ learning curve. I can recommend two Symbian C++ IDEs:

- The Borland C++BuilderX Mobile Edition is available at *http://www. borland.com/mobile/cbuilderx/index.html*.

- The CodeWarrior Development Studio for Symbian OS is Nokia's Symbian C++ development tool. You can download and purchase it from the tools section in the Forum Nokia web site (*http://www.forum.nokia.com/*).

To learn more about Symbian C++ development, please refer to the book *Developing Series 60 Applications* (Addison Wesley, 2004).

 ## Run Python Scripts

#18   Use the Python scripting language to develop small hacks for your Series 60 device.

While Java and Symbian C++ are powerful programming languages, their learning curves are too steep for most smartphone users. Most users do not need a full-blown programming platform to develop small hacks for their phone. A scripting language is the perfect tool to automate simple tasks and perform simple logical processing.

Python is a widely used scripting language in the computer world. It is easy to learn and supports object-oriented program construction. Nokia provides support for Python on most of its Series 60 devices.

 Speaking of running scripting languages on Nokia Series 60 smartphones, there is an unsupported Symbian port for Perl 5.8x and 5.9.x. You can find the installation package and usage instructions from this mailing list posting from a Nokia engineer: *http://www.xray.mpe.mpg.de/mailing-lists/ perl5-porters/2005-04/msg00439.html*.

Install the Python Environment

The current Nokia phones do not come with the Python runtime environment preinstalled. You have to download and install Python yourself.

You can download the Python for Series 60 package from the Forum Nokia web site under the Series 60 Platform → Tools and SDKs category. The download package is a zip file with the *.sis* installation files, documentation, and example code. Make sure you read the Getting Started document in the download bundle to choose the correct *.sis* file for your phone. Then, you need to install the extracted *.sis* file to the phone, following the instructions in "Run Symbian Applications" **[Hack #17]**. The *.sis* file installs the following components to the phone:

- A Python language interpreter
- The necessary libraries (DLL files) to run Python applications
- A plug-in for the phone to recognize Python scripts and Python libraries downloaded from the Internet or embedded in incoming messages

Once the Python runtime is successfully installed, a Python icon appears in the Main menu.

Run Python Scripts

Click the Python icon to open it and then select Options → Run script. You should see a list of installed Python scripts and applications (see Figure 3-10). If this is the first time you've run Python, the scripts that appear will be the demo scripts from Nokia. You can try any of them. The *filebrowser.py* script is a good start. It allows you to browse the filesystem on the Series 60 smartphone, similar to what FExplorer does **[Hack #20]**).

Install Python Scripts

Of course, ultimately you want to install and run your own Python scripts on the device. To do that, you can put the Python script (with the *.py*

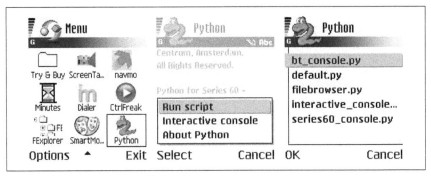

Figure 3-10. Running an installed Python script

filename suffix) on a web page and download it using the Services browser, or you can simply send the script to the phone via Bluetooth. The phone recognizes the Python script and prompts you to install it. Once the script is installed, it becomes available under the Python script list and you can run it from there. Figure 3-11 shows the process.

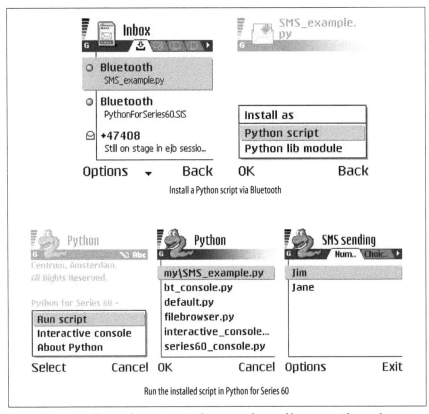

Figure 3-11. Installing and running a Python script from a file sent via Bluetooth

The user-installed scripts are placed in the *C:\System\Apps\Python\my* directory (or in *E:\System\Apps\Python\my* if Python for Series 60 is installed on the MMC card; see Figure 3-12). You can delete those files to delete Python scripts.

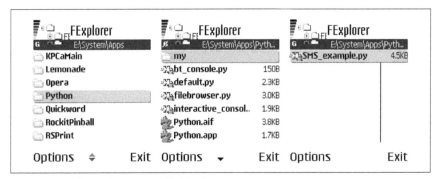

Figure 3-12. The path to an installed Python script

Write Your Own Python Scripts

You can write your own Python scripts in any PC text editor and deploy them to the phone using the web or Bluetooth methods described earlier. But the key benefit of a scripting language is the ability to quickly prototype scripts. In fact, Python for Series 60 allows you to prototype Python scripts right there on the phone!

You can use the *interactive_console.py* script that comes with the Python runtime to open a text console for the Python interpreter (or, you can simply use the Options → Interactive console soft-key menu). Then, you can type Python statements from the phone keypad and see them run! Figure 3-13 shows the console executing several simple Python statements. The print statement and math formula statement generate output in the console. The statements at the bottom of the last image in Figure 3-13 display a Series 60–style UI dialog box.

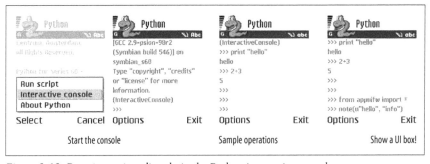

Figure 3-13. Running scripts directly in the Python interactive console

Unlike full-blown programming languages such as Java and C++, which require setup code and noncode resources for even the simplest application, a scripting language such as Python actually allows you to write useful programs with only several lines of code. A great way to learn Python is through examples. The BigBold web community has a page for user-posted Series 60 Python examples (*http://www.bigbold.com/snippets/tags/series60*). There, developer Korakot Chaovavanich posted several useful snippets covering topics such as camera operation, Short Message Service (SMS) messaging, file exchange, and SQL database queries. Check them out, and share your own!

Write Series 60 Python Scripts on a Computer

Programming Python with the mobile-phone keypad is not pleasant. Luckily, the *bt_console.py* script bundled with Python for Series 60 opens a Bluetooth console on the device, which can be connected to a console program on a computer. The computer console sends everything it receives from the keyboard to the device via a serial port emulated over the Bluetooth connection. In turn, the phone console sends all the response data back to the computer console. This allows you to type Python commands on a regular computer keyboard and have them executed on the phone in real time. Very cool!

Series 60 Python scripts do not run in the regular Python environment on a desktop computer, since they use Nokia-specific Python libraries for UI, messaging, and multimedia operations. The Series 60 Compatibility Library (*http://pdis. hiit.fi/pdis/download/*) ports the phone library to computers and allows you to run Series 60 Python scripts on Windows, Mac OS X, and Linux computers. This library is developed by the Personal Distributed Information Store project at the Helsinki Institute for Information Technology.

On a Windows PC, you can use the console program bundled in the Nokia Python SDK to work with the Bluetooth Python console on the phone. Read the Nokia Python SDK documentation to see how it works.

The Nokia documentation does not cover how to run the serial console on Mac or Linux computers. But it is a straightforward process once you set up the Bluetooth serial port on the computer. Here are brief instructions for Mac computers, adapted from Erik Smartt's blog at *http://www.eriksmartt.com/ blog/*:

1. Pair the phone with the Mac computer (see "Configure Bluetooth for Mac OS X" [Hack #13]).

2. Set up an incoming RS-232 port over Bluetooth using the Bluetooth Serial Utility program in the Applications → Utilities folder. Give the serial port a name (e.g., PythonConsole) and remember it.

3. From a Terminal window (Applications → Utilities → Terminal), you can use the screen /dev/tty.*portname* command to redirect that port to the terminal. *portname* is the name of the port you assigned in the last step (i.e., PythonConsole).

4. On the phone, use the BTConsole application to search for available Bluetooth devices and choose your Mac from the list. You should see a Connected message in your Mac Terminal window, followed by some directions and a Python prompt.

After you are finished with console programming, you can press Ctrl-D on the Mac to exit the BTConsole and shut down the process on the phone.

Quick Access to Applications

You might install a lot of applications, and if you're on a Series 60 phone, you might run a bunch of them at once. With all this potential disorder, you need to be able to quickly launch and switch between applications.

Nokia smartphones can be loaded with applications—both factory-installed and user-installed. However, to launch any application, you must go through multiple menu items (for Series 40 devices) or scroll up and down in a grid (for Series 60 devices). It is a slow process for busy people on the move.

Different phone users want quick access to different applications. For instance, a mobile photographer probably wants to start the Camera application quickly to catch a precious moment. The frequent instant-messaging user wants to keep the Messaging application available at all times, even though he might temporarily switch to other applications from time to time.

Assign Soft-Key Shortcuts

The easiest way to add a shortcut to an application is to assign it to a soft key in the phone's idle screen (a.k.a. the home screen). On a Series 60 device, you can use the Tools → Settings menu and then select Phone → Standby mode to customize the soft keys (see Figure 3-14). On a Series 40 device, the menu path is Settings → Personal shortcuts (see Figure 3-15).

Some Series 40 devices, such as the Nokia 6230, allow you to customize the application shortcut for the right soft key only. The left soft key is always mapped to the GoTo menu, which contains a customizable list of applications and bookmarks for quick access, and the middle soft key is always mapped to the Main menu.

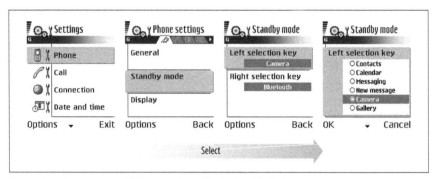

Figure 3-14. Assigning soft-key shortcuts for Series 60 devices

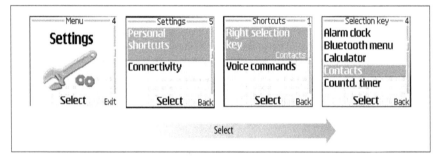

Figure 3-15. Assigning soft-key shortcuts for Series 40 devices

Notice that only the factory-installed applications can be assigned to the idle-screen soft keys.

 On some Series 60 phones, such as the Nokia 3650, assigning the Call Forwarding application to a soft key is the only way you can launch the application.

Speed Launcher

If you need quick access to a user-installed application on a Nokia Series 60 device, you can use the speed launch feature, which allows you to launch an application with just one touch of the keypad from the Main menu. To do that, first you press the menu key to enter the Main menu. All the applications and top-level folders are listed in a grid of icons in this menu. For the first nine applications in the grid, you can simply press their corresponding number key to launch them. For instance, in Figure 3-16, the Opera browser (it's the "O" with a "www" underneath it) is in the position corresponding to the number 3 key on the keypad—that is, the rightmost key on the top row. If you press the 3 key, the Opera Mobile Browser for Series 60 application is automatically launched. Please note that the numerical shortcut works only immediately after you enter the Main menu. If you use the navigation pad (a.k.a. joystick) to move the highlight around, the shortcuts stop working. They will work again if you go back to the idle screen and get back into the Main menu.

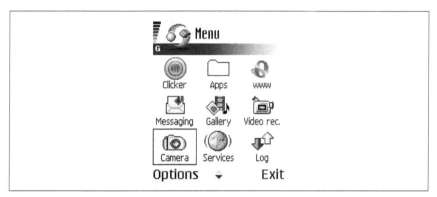

Figure 3-16. One-touch launch for applications in this grid

 The 0 key does not launch any application in the main menu screen. But in the idle screen, if you press and hold the 0 key, the phone's default web browser (i.e., the Services application) is automatically launched.

You can use the Options → Move menu to move the icons around in the grid (see Figure 3-17), and hence, change their speed launch numbers. Naturally, the application icons near the top of the grid are easier to access since they require less searching and scrolling once you are in the Main menu.

Quickly Switch Applications

The slow response of launching applications is one of the major complaints among Series 60 device users. As most users have several frequently used applications, a neat trick is to keep those applications running in the background all the time, thereby saving on startup and shutdown time.

 The background applications are possible only with the multitasking Symbian operating system. This feature is not available on Series 40 devices.

You can place a running Series 60 application in the background by pressing the red "End call" key (the key with the red telephone symbol) while the application is running. The next time you open the same application, it will pop up instantly and appear in the state exactly as you left it.

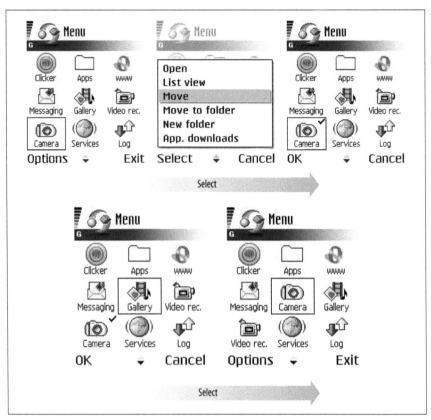

Figure 3-17. Moving the icons in the grid to optimize access

You can see a list of background applications by pressing and holding the Menu key (see Figure 3-18). You can select any application from the list and bring it to the foreground. If you press the c key while you are scrolling through the list, you can force the background program to exit.

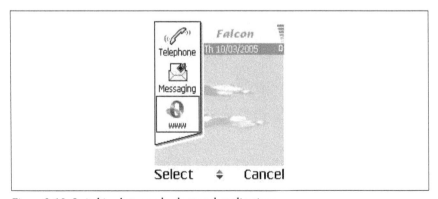

Figure 3-18. Switching between background applications

Voice Commands

Finally, most Nokia smartphones have a utility application for voice commands. Depending on your phone model, you can access it via the Tools → Voice commands menu, or the Extras → Voice commands menu, or the Settings → Personal shortcuts → Voice commands menu. Using the voice command utility, you can map any voice recording to a selected number of applications. For instance, you can record yourself speaking the word *camera* and map it to the Camera application. The voice commands for applications work similar to the voice dialing tags [Hack #26]. You simply press and hold the right soft key for the phone to pick your spoken command and launch the matching application.

Explore the Filesystem

Nokia devices feature PC-like filesystems. The Gallery application supports limited access to media files. But using special software, you can get around these limits and do a lot of cool things with the files.

With all the talk about how a smartphone is supposed to have PC-like functionality, it still does not quite feel like a PC. On a PC, the data and applications are separate; the data is stored in files in the PC's hard drive, and you can use any application to manipulate any datafile in the system. For example, you can use a text editor instead of the email program to open an email message and see what's going on inside. That gives you the flexibility to determine how to use your data.

A mobile phone, on the other hand, integrates the data with its handling application. For instance, the email messages are accessible only from the Messaging application and the contacts are accessible only from the Contacts application. However, under the hood, the phone still organizes data and executable programs into files and folders. In this hack, I'll cover the concepts of the mobile phone filesystem, the tools you can use to explore it, and what you can do with the raw files.

The Native Gallery

On Nokia devices, the closest thing to the file explorer is the Gallery, which is available on both Series 60 and Series 40 devices. Inside the Gallery, you can choose to access either the phone's main memory storage space or the MMC card. In each storage unit are several folders with names such as *Images*, *Sound clips*, and *Video clips*. Those folders hold files and subfolders. They function just like folders on a computer. When you open a file in the Gallery, the phone operating system uses the filename suffix to determine which application should be used to open any particular file. For instance, a

.jpg file is opened with an image viewer; a *.3gp* file is opened with a video player (e.g., the RealOne player), etc. You can rename, delete, and move files around from one folder to another.

> On older Series 60 devices (e.g., Nokia 3650), there is no Gallery application. On those devices, the Images application is equivalent to the Gallery found on newer Series 60 and Series 40 devices. Despite its name, the Images application holds images, audio files, video clips, and other multimedia files.

The limitation of the Gallery, of course, is that you have no control over which files are saved in it. The phone automatically saves the following media files into the Gallery:

- Image files from photos captured from the Camera application
- Audio files captured from the Recorder application
- Video files captured from the Video Recorder application
- All media files of known formats downloaded via the Services browser
- All media files of known formats downloaded from the Messages Inbox

Nonmultimedia files, such as installation package files (i.e., *.sis* and *.jar* files), text files, office documents, and executable files, cannot be saved into the Gallery.

Introducing FExplorer

To gain full access to the phone filesystem, you need special software to get around the limitations imposed by the phone UI. For Series 60 devices, the freeware FExplorer, written by Dominique Hugo, does that for you. You can download the latest version of FExplorer, or make a donation to support its development, at *http://www.gosymbian.com/*. Alternatively, a commercial product called Extended File Manager, from Psiloc (*http://www.psiloc.com/index.html?id=159*), has similar functionality.

> What about Series 40 devices, you ask? Series 40 devices use the closed Nokia OS. Hence, only native applications developed by Nokia can have direct access to the filesystem. Nokia has not released any such application. In the future, the Java runtime on Series 40 devices might support the J2ME File Connection extension API, which would support file access from third-party Java applications. But for now, there is no reliable way for you to access the underlying filesystem on Nokia Series 40 devices.

The filesystem naming conventions that the Symbian OS uses are similar to those in the Windows OS. The *C:* drive letter is for the storage space for the system-level files in the phone's internal flash memory; *D:* is the RAM disk that stores runtime information; *E:* is the MMC card, storing applications and media files; and *Z:* is the Read-Only Memory (ROM) for the device's system software. The FExplorer program lets you create, delete, and manipulate directories and files in the filesystem. It also allows you to send any file to remote devices via Bluetooth, IR, MMS, or email. Now, let's explore some common files on a Nokia Series 60 smartphone.

Media files. The *C:\Nokia* and *E:* directories contain all the top-level folders in the Phone Memory and MMC Card tabs of the Gallery, respectively (see Figure 3-19).

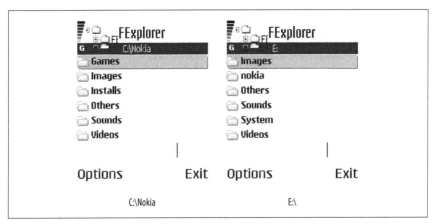

Figure 3-19. The media Gallery directories in the FExplorer program

So, why would anyone use FExplorer rather than the native Gallery to manipulate those media files? Well, for a couple of reasons:

- FExplorer supports file search via the Options → File → Find menu. You can search the entire device or any directory using PC-style wildcards in filenames (e.g., scr*.jpg).

- It is easier to move files to other directories using the Options → Edit → Copy/Paste menu in FExplorer than it is with the file move function in the Gallery.

- FExplorer allows you to move nonmedia files into and out of any folders on the device. I will discuss this point in more detail in the next section.

My Nokia 6600 device has a File Manager application in the *Extras* folder in the Main menu. But it does not provide any capabilities beyond the Gallery application.

Files received in the Messaging inbox. In the *C:\System\Mail* directory, you can find all the messages you've received. The MMS, Bluetooth, IR, and email message attachments are stored as files in the nested subdirectories (see Figure 3-20). The *Mail* directory might contain many cryptic subdirectories. The best way to locate a particular attachment file is to search for its name via the Options → File → Find menu.

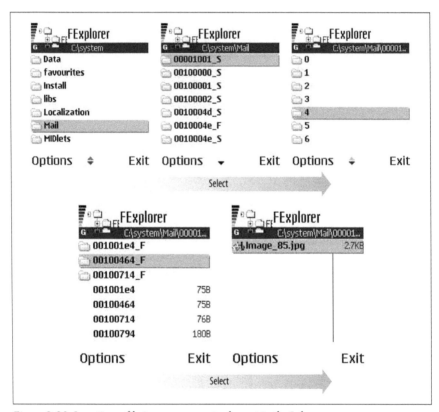

Figure 3-20. Locating a file in a message attachment in the Inbox

This feature is especially useful when you need to access a received file that the device cannot recognize. For instance, suppose someone sends a PDF file to your phone via Bluetooth. If you have not installed a PDF reader application, the phone will not recognize this file and will not offer you an option to save it to the Gallery. But with FExplorer, you can locate the file and then send it via Bluetooth to a PC where you can read it. Or, you can copy it to the Gallery. If you install a PDF viewer on your device **[Hack #39]**, you will be able to read it.

Installation packages. The installed Java MIDlets are located in the *C:\ System\MIDlets* and *E:\System\MIDlets* directories. You can find the *.jad* and *.jar* file pair for each installed MIDlet (see Figure 3-21), and you can open the *.jad* file and read its contents. The *rms.db* file contains the persistent storage data this MIDlet stores on this device.

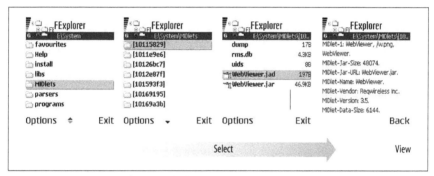

Figure 3-21. The MIDlet files installed on the device

For each installed Symbian application, the *.sis* installer is cached in the *C:\ System\Install* or *E:\System\Install* directory (see Figure 3-22). The installed executable files and runtime configuration files are located in the *C:\System\ Apps* or *E:\System\Apps* directory (see Figure 3-23).

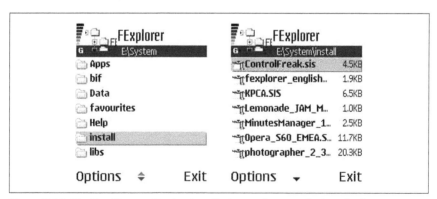

Figure 3-22. The Symbian application installation packages cached on the device

You can use FExplorer to send the *.sis*, *.jad*, and *.jar* files to other devices. But make sure you have the proper rights to do so. On some devices, the built-in Digital Rights Management (DRM) system does not allow you to

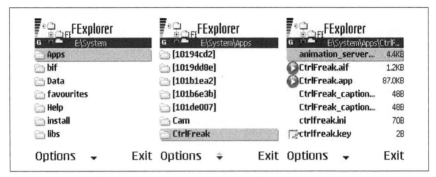

Figure 3-23. Installed executable and configuration files for Symbian applications

send out files with those suffixes. You can get around this by simply changing the filename to something else and then changing it back after sending.

Other cool features. In addition to the file explorer, FExplorer supports some other cool features that you might want to check out. You can access all of those features via the Options menu (see Figure 3-24).

- It allows you to take screenshots of the phone and save them to a specified directory in the Gallery.
- It can compress the memory to increase available storage space (i.e., defragmentation [Hack #21]).
- It allows you to set and remove operator logos using appropriate image files [Hack #46].
- It displays the phone information including model number, IMEI number, battery status, and software version number.
- It displays network information including the ID of your current cell base station and your service provider name.

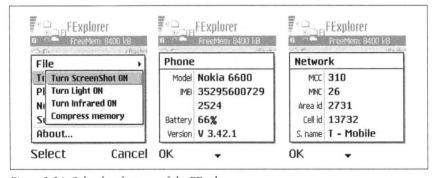

Figure 3-24. Other key features of the FExplorer program

The FExplorer program is a very useful tool, and I will use it again several times later in this book.

 ## Manage Your Phone's Memory

Store more pictures and improve the device's performance with these memory management techniques.

A typical Nokia smartphone has between 100 KB and 12 MB of built-in memory. This memory holds everything from applications and media files to messages and dynamic data created by applications at runtime. Optimizing how this memory is used will let you store more data, run more applications, and improve the general performance of the phone. In this hack, I'll check out some cool memory management tips.

Use the Memory Card

To expand the memory space available on the device, you can install a flash-based MMC card. The size of an MMC card can vary from 32 MB to 2 GB—depending on your encoding settings, 512 MB and larger cards should be sufficient for hours of movies or digital music (see "Play PC Video Clips on the Phone" [Hack #71], "Play DVD Movies on the Phone" [Hack #72], and "Play Digital Music" [Hack #74] for information on how to play video and music on your phone). Most Nokia phones, such as the Nokia 6230, 3650, 6600, and 6620 smartphone, support the standard MMC cards you can buy from any electronics store. Newer Nokia phones, such as the Nokia 6630, 6670, 6681, and 6682 phones, support only a Nokia proprietary MMC format known as Dual Volt Reduced Size MMC (DV-RSMMC). The DV-RSMMC card is slightly smaller than the regular MMC card, but the former is more difficult to find and is more expensive. You should read your owner's manual or the phone specification on Nokia web sites to find out exactly what MMC card format your phone supports.

To install the MMC card, you can just insert it into a socket underneath the phone battery. For some devices, such as the N-Gage QD, the MMC card socket is located in an external card slot for easy access. Refer to your phone manual for the exact installation steps. Once the MMC card is installed, you can inspect its properties via the Extras → Memory menu on a Series 60 device or the Gallery → Memory card → Options → Details menu on a Series 40 device.

 On a Series 40 device, you can view the status for both the phone's main memory and the MMC card via the Settings → Phone settings → Memory status menu.

Installing the memory card is the first step. To use the card, you should configure applications to store data on it. In the following list, I provide instructions for installing some key applications on a Series 60 device. Series 40 devices work similarly.

Gallery

The Gallery application now displays folders in both the phone main memory and the memory card. You can move files between the two storage units.

Application installers

When you install Java or Symbian applications, the installer asks you whether to put the application on the memory card. Please note that applications installed on a memory card still leave a small stub in the main memory. Also, the applications will stop working if the memory card is removed.

Camera, video recorder, and audio recorder

In the Settings menu of the Camera, Video Recorder, and Audio Recorder applications, you can choose "Memory in use," located between "Phone memory" and "Memory card." The captured media files are automatically saved to the appropriate folders in the selected storage unit.

Messages

In the Messaging application's Settings → Other menu, you can choose "Memory in use" for saved messages (see Figure 3-25). If you choose "Memory card" here, the phone saves all received SMS, MMS, email, and Bluetooth messages to the memory card. This feature is available only on Series 60 devices.

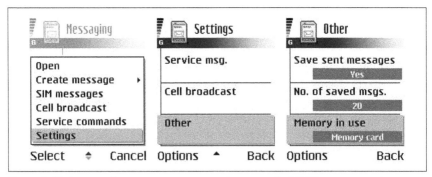

Figure 3-25. Storing all incoming messages and songs to the MMC memory card

The memory card is a little slower than the built-in phone memory. The memory card stores persistent information. It does not store the dynamic data generated by applications at runtime.

> If you use Bluetooth to send any file larger than 4 MB to the phone (this limit is lower on some older phones), you have to set the Messaging application's memory in use to "Memory card." If you do not do that, the message-receiving operation will be aborted by a "Memory full" error because the phone's main memory is only a couple of MB in size and cannot hold the received file.

Reduce Logging

Nokia phones can log all calls and network connections it initiated and received over a period of time. The default logging period is the past 30 days. While this is a very useful feature, the logging data can consume a lot of memory and slow down the device, especially if you make and receive a lot of phone calls.

You can reduce the size of the log by changing the Options → Settings → Log duration value in the Log application on a Series 60 device, or you can simply delete all log entries from the Log application when you need more memory space. On a Nokia Series 40 device, you cannot configure the log duration but you can delete log entries via the Call register → Delete recent call list menu item.

Reduce Concurrent Applications

As discussed in "Quick Access to Applications" **[Hack #19]**, Series 60 devices can run multiple applications at the same time. Each application uses some memory to store its runtime data, and uses some CPU cycles. So, if you have too many concurrent applications, the phone slows down noticeably and starts to throw memory-related errors. If that happens, you should try to exit some of the applications.

Memory Management Software

If you are really tight on memory, you can try some of the memory compression software solutions for Series 60 phones. The FExplorer tool discussed in "Explore the Filesystem" **[Hack #20]** can compress memory by moving fragmented storage units into a contiguous block. This operation typically increases available memory by 10% (see Figure 3-26).

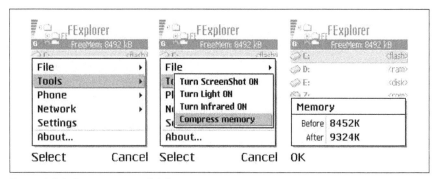

Figure 3-26. Compressing and defragmenting memory

If the FExplorer memory compressor is still not good enough, Symbian software tools are available for compressing in-memory data and applications by up to 60% when they are not in use. These memory compression tools can expand the compressed files when they are needed. They essentially trade the persistent storage space for CPU processing cycles. Two popular data compression programs for Nokia Series 60 devices are:

ZipMan

> ZipMan is a zip archive utility for Series 60 devices. You can build compressed archives of images, video clips, and other files from the Gallery. When you need to use those files later, ZipMan extracts them for you. You can download and purchase ZipMan from *http://www. wildpalm.co.uk/ZipMan7650.html*.

Space Doubler

> Space Doubler compresses Symbian applications. It runs in the background to expand the compressed application on the fly whenever you try to execute the application. From the user's point of view, the compressed application behaves exactly the same way as the regular applications, except that the startup time is slower due to the extra expansion process. You can download and purchase Space Doubler from the following web site: *http://www.psiloc.com/index.html?id=154*.

In general, I do not recommend using these two solutions on a regular basis, unless you are desperate for space, since they slow down the device noticeably.

Reset and Restore

The phone's memory consumption tends to increase and the performance tends to decrease over time, especially if you install and uninstall programs frequently. This is because the phone OS cannot efficiently reuse the fragmented memory space freed by the uninstalled program or deleted datafiles. You can improve your phone's memory performance by doing a deep reset to clean up the memory and then reinstalling all the programs and datafiles. Please refer to "Reset and Restore Your Phone" **[Hack #23]** for more details.

Protect Your Phone

Hacks 22–25

In the mobile information age, mobile phones not only store sensitive information about us, but also act as an identity provider for us when we use mobile network services. Therefore, it is important that we keep our phones physically secure.

It is also important that we keep our phones secure from viruses and other forms of attack. As mobile phones and related services become increasingly sophisticated, many potential security problems can arise because phones are networked in so many ways. In this chapter, you will learn how to secure your device against malicious programs, service theft, and data theft.

HACK #22 Avoid Malicious Software

Protect your Nokia phone against mobile phone viruses, malware, Trojan horses, Bluetooth scanners, and other malicious programs.

As a powerful and connected computing device, the Nokia smartphone suffers the same vulnerability as other connected computers—viruses and other malicious programs can attack them over the network. Those programs can potentially harm the device, leak sensitive information, cause data loss, or even result in big service charges for you. Compared with regular computer viruses, a mobile phone virus can be especially harmful, since it can spread in peer-to-peer local networks; worse, most phone users are not prepared to deal with such viruses.

Basics of Malicious Programs

Before you can act to protect yourself, you need to know a little about how malicious programs can attack your mobile phone. The following is a list of representative malicious programs on smartphones and the harm they cause:

Force the phone to perform operations that interfere with regular user operations
The original Cabir virus (*http://www.sarc.com/avcenter/venc/data/epoc.cabir.html*) forces your phone to scan the Bluetooth network all the time, quickly draining the battery.

Disable some device functionality
The Dampig virus (*http://www.sarc.com/avcenter/venc/data/symbos.dampig.a.html*) replaces some key system libraries and makes many applications, including the Bluetooth user interface (UI), useless on your phone. The Locknut virus (*http://www.sarc.com/avcenter/venc/data/symbos.locknut.html*) can cripple your phone to the point that you cannot make voice calls. The Fontal.A virus (*http://www.sarc.com/avcenter/venc/data/symbos.fontal.a.html*) does not cause immediate problems for your phone, but it does secretly replace several key font files, which prevents the phone from booting up once you reboot it. Bluetooth scanners can send malformed Bluetooth messages to your phone and crash its Bluetooth program, forcing you to reboot your phone to recover.

Make phone calls or send Short Message Service (SMS) messages to expensive caller-paid services
The Mosquitos Trojan (*http://www.sarc.com/avcenter/venc/data/trojan.mos.html*) sends messages to premium SMS service numbers without your knowledge or approval. The message costs are billed directly to your service account. Some Bluetooth-based exploits allow a Bluetooth scanner running on a nearby device to remotely dial your phone or make arbitrary connections using AT commands.

Leak out sensitive personal information
Malicious Bluetooth scanners can allow a cracker to remotely steal the entire memory contents of your phone from another nearby device without your knowledge. In theory, it is also easy to develop a mobile Trojan that gathers information about your contacts, calendar, and media Gallery, and then sends the information to a third-party server on the Internet.

 Cabir is the first virus known to target Nokia Series 60 devices. It is largely a proof-of-concept virus. Cabir spreads over Bluetooth and does not contain a *payload* (the malicious software that does the actual harm). It is benign, except for the fact that it drains your battery with continuous Bluetooth searches. Later variations of the Cabir virus, such as Cabir.b and Lasco, can do real harm to you and your phone.

Based on their attack methods, malicious software on smartphones can be divided into two categories:

- Mobile virus or Trojans that are downloaded and installed into your smartphone
- Bluetooth scanners that remotely exploit your phone from another nearby device

Now, let's discuss those two types of attacks in more detail.

Viruses and Trojans. Currently, all Nokia mobile phone viruses are written in Symbian C++ and are deployed to devices as Symbian programs. Although in theory Java-based viruses are possible, they are substantially more difficult to develop and deploy, since Java applications must run in the Java Virtual Machine and must conform to strict Java security policies. Since Java applications do not have direct access to your phone's physical memory or other low-level device-native features, it is less likely that they can breach or circumvent the phone's security policies. In fact, there is no known Java virus for Nokia phones. Since Java is the only programming platform on Nokia Series 40 devices, there are no known Nokia Series 40 viruses.

Mobile viruses and Trojans must be downloaded into your phone for them to take effect. Viruses and Trojans can spread in three primary ways:

Trojan download

The malicious program can present itself as a known (or appealing) Symbian program and trick you into downloading and installing it directly. For instance, the Mosquitos Trojan virus poses as a *cracked* version of the popular Symbian game, Mosquitos, on certain file-sharing networks. A cracked version of a game is a version that's been illegally modified to remove the registration module, so you can play it for free. The idea is that you'll run it, thinking you are running a game, but the Trojan virus will activate when you run it. Other examples include the Dampig virus, which pretends to a cracked version of the FSCaller application, and the Skulls virus, which pretends to be a theme manager application. To prevent Trojan viruses, you just need to be careful about the sources of the programs you download. I recommend that you use only legitimate software downloaded from well-known web sites. Beyond the immediate concern of security, it also helps if you don't try to circumvent copy protection, and instead, support the developers that work hard on software you want to use.

Bluetooth

> Viruses can spread over the local Bluetooth network. An infected device tries to find all Bluetooth devices in its neighborhood, all the time. Once a device is found, the infected device sends the program over to the new device. The recipient is then presented with a message to accept the incoming file and install it. The original Cabir virus spread in this way. If the recipient is not well informed or if the message is deceiving, he might just install the program. For instance, the Gavno virus presents itself as a "software patch," borrowing a familiar concept from Microsoft Windows to deceive users. Once the program is installed, it can execute itself and then start to search for nearby Bluetooth devices to spread further.

MMS

> A Bluetooth-based mobile virus can infect devices only within a range of several meters. Hence, the virus can travel only as fast as the devices move, which is the speed of airplanes in modern societies. Some newer mobile phone viruses, such as the Commwarrior, can spread over MMS. The virus tries to send itself via MMS to 256 random phone numbers from your Contacts list. This can potentially allow the virus to spread at the speed of telecommunications, which means it can spread across the world in a very short period of time. And what do you do when you receive an MMS from a friend? You open it, of course. This is the same kind of social engineering that permitted so many Microsoft Outlook-based viruses to spread over the years.

> Some Nokia devices' Bluetooth implementations have known security vulnerabilities that allow files to be received without user acknowledgment. If this vulnerability is exploited by a Bluetooth-based virus, it can be extremely dangerous.

Bluetooth scanners. Bluetooth scanners exploit insecure implementations of the Bluetooth system software on some phone models. Several Nokia phone models are known to be vulnerable (e.g., Nokia 7650, 6310i, etc.). You can get more information, including an updated vulnerable-device list, from *http://www.thebunker.net/security/bluetooth.htm*.

> Bluetooth exploits were first discovered by Adam Laurie, of A.L. Digital Ltd., in 2003.

A Bluetooth scanner has to be physically close to your phone (e.g., in a conference hall or classroom) for Bluetooth to work. There are three known types of Bluetooth attacks:

Bluesnarf

> This type of attack can be launched from untrusted (a.k.a. unpaired [Hack #11]) devices. The attacker can steal information, including your Contacts list, calendar, photos, etc., from your phone.

Backdoor

> This type of attack has to be launched from a previously paired device. The attacker can get access to almost all the functionality on your phone.

Bluebug

> This type of attack involves creating a Bluetooth serial profile [Hack #11] to your phone, and then hijacking the phone's voice and data connections.

> Bluejacking is often cited as a fourth type of Bluetooth attack. But it is really just a prank. It works as follows. The prankster creates a contact entry on her own phone and enters a prank message into the "name" field. For instance, the "name" of this contact might be "Your phone belongs to us." Then, the prankster sends the contact to random Bluetooth phones as a business card [Hack #35]. The recipient suddenly sees an unsolicited prank message—"Your phone belongs to us"—on his phone screen.

Preventive Measures

The best protection is prevention: knowing how the malicious programs work. You can take several simple precautions to minimize your risk.

Only install trusted programs. The key to prevent viruses and Trojans is to be extremely careful about what you install. If you download *.sis* applications from the Web, you need to verify that they are indeed legal and that they come from an authorized source. If you receive an application over Bluetooth, as a general rule do not install it unless you already had a conversation with the sender and are expecting it. Do not install *any* program from email or MMS message attachments.

Minimize Bluetooth exposure. To minimize the risk of Bluetooth-based viruses and Bluetooth scanners, you should turn off Bluetooth in public places. If that is not possible, you can make the device invisible (see Figure 4-1) so that it does not show up when other devices scan the network. If you do receive a Bluetooth message from a friend, talk to her and confirm her intentions before you accept or install the application file.

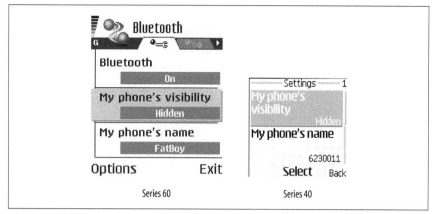

Figure 4-1. Making the phone invisible on Bluetooth networks (for both Series 60 and Series 40 devices)

Use a personal firewall. Most malicious mobile programs rely on the network to spread or work. You can prevent them by controlling the network connections on your phone. One of the most effective network control tools is a firewall. By installing a firewall on your phone, you can:

- Prevent unauthorized Bluetooth or General Packet Radio Service (GPRS) incoming connections and file transfers
- Prevent Bluetooth scanners from discovering or pairing with your phone
- Prevent Bluetooth scanners from accessing any data or services on your phone
- Prevent Trojans from sending out any information from your phone

The Symantec Mobile Security for Symbian (currently in beta) provides a personal firewall for Nokia Series 60 phones. You can download it from *https://www-secure.symantec.com/public_beta/*. Via the firewall, you can specify several different levels of communication constraints. Figure 4-2 shows the mobile phone firewall in action.

Remove the Virus

If you know the name of the virus that infected your device, you can search for it via Google. You can probably find a lot of security bulletins from research sites such as *http://www.symantec.com* and *http://www.f-secure.com*. Most of these bulletins include a complete description of the virus, including the files it installs on your device. For instance, the following two URLs point to the F-Secure and Symantec bulletins for the Cabir virus:

- *http://www.f-secure.com/v-descs/cabir.shtml*
- *http://securityresponse.symantec.com/avcenter/venc/data/epoc.cabir.html*

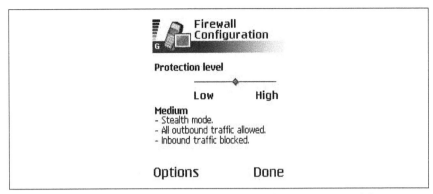

Figure 4-2. Mobile firewall on a Nokia Series 60 device from Symantec Mobile Security for Symbian

With a file browser tool such as FExplorer [Hack #20], you can follow the instructions to remove the virus from your device.

In practice, it is difficult to know the exact name of a virus. For instance, the Cabir virus has at least eight very similar variations. So, the preceding method is not always practical in the real world. In most cases, a much simpler way to erase the malicious program and reverse the damage is to perform a deep reset. "Reset and Restore Your Phone" [Hack #23] discusses how to reset your phone and then restore its functionality via data backups.

> If your phone has been infected with a virus and you do not know which programs are infected, it is probably a good idea to be conservative and install all third-party programs from scratch instead of simply reloading them from the backup.

Use Antivirus Software

In the previous two sections, I covered generic approaches to protect your phone and recover from an attack. But those approaches do not always prevent all attacks. Manually resetting the phone is time consuming and does not reverse the financial loss you might have incurred from the virus.

If you are really concerned about mobile viruses, you can invest in antivirus software to protect your phone. Antivirus programs for Symbian-based phones (e.g., the Nokia Series 60 smartphones) are available from the following vendors:

- F-Secure Mobile Anti-Virus from *http://www.f-secure.com/products/ fsmavs60/*

- SimWorks Anti-Virus from *http://www.simworks.biz/sav/*

- Symantec Mobile Security for Symbian (beta) from *https://www-secure. symantec.com/public_beta/*

Like antivirus software on computers, mobile antivirus software scans all files on your device to look for specific patterns of known viruses (a.k.a. the virus signature). If it finds one, it isolates the infected file and presents you with the option to remove it. Figure 4-3 shows a full scan performed by the Symantec Mobile Security for Symbian program (beta).

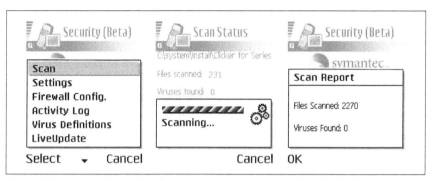

Figure 4-3. Scanning the entire phone using Symantec Mobile Security for Symbian

The full device scan takes a long time and consumes a lot of battery power, so don't perform a full scan on a regular basis. To save time and energy, the antivirus programs support incremental scan modes that check only incoming files, such as files you've downloaded or created using software on your phone. After the first full device scan, the antivirus programs run in the background and automatically scan all incoming files from the Web, Bluetooth, MMS, and email messages, and check for virus signatures as they arrive. If they detect an infected program, you will be advised not to install it.

As you would expect, the key for a successful antivirus program is to have a complete list of virus signatures to check against. This is a moving target, since new viruses might be written after the antivirus software is released. So, all mobile antivirus programs come with a subscription service that allows the program to update its virus signature database periodically over the Internet via the phone's data connection.

Reset and Restore Your Phone
Reset your phone without losing data or custom settings.

As a power user, you often need to fiddle with your phone to try out new software, experiment with new configuration settings, or even clear up viruses or other malicious programs. This fiddling can sometimes cause the phone software to crash, freeze, or otherwise behave abnormally (e.g., it will be unable to connect to the network or unable to run some programs). Some phone viruses discussed in "Avoid Malicious Software" **[Hack #22]** can also freeze your phone. This is when the reset feature in Nokia phones comes in handy. In this hack, you will learn not only how to reset your phone, but also how to back up and restore data in a systematic manner.

Two Types of Resets

Nokia phones can handle two types of resets: normal reset and deep reset.

Normal reset. On a Series 40 device, you can do a normal reset via the Settings → Restore factory settings menu. The normal reset simply resets the most basic phone settings, such as the security code, whether to enable speed dialing, the display brightness, and the screensaver time-out, to their factory preset values. It does not alter the applications mapped to the right soft key (i.e., the GoTo key), nor does it alter any of the network connection settings.

On a Series 60 device, you can do a normal reset by entering the service code *#7780# on the phone's idle screen, as though you are making a phone call. Alternatively, you can use the Tools → Settings → Phone settings → General → Orig. factory settings menu to perform the normal reset.

The Series 60 normal reset does everything the Series 40 normal reset does. In addition, it forces the phone to reload all the .*ini* files from the Read-Only Memory (ROM) to the C: drive under the Symbian OS. Hence, many application settings are restored. For instance, a normal reset on a Series 60 phone restores the soft-key shortcuts on the idle screen to factory settings, deletes the phone's Bluetooth name, eliminates all the GPRS access point settings **[Hack #10]**, and resets data storage options in the Camera or Messaging application to "Phone memory" **[Hack #21]**. The email Inbox settings **[Hack #60]**, however, are left untouched.

 The phone prompts you for the current security code before it resets itself. The default security code is 12345.

On both Series 40 and Series 60 phones, the normal reset leaves intact all the user data on the phone, such as contacts, calendar items, wallpapers, images, tones, messages, and third-party applications. While it is a safe operation, the normal reset is of limited value, since most of the time it is the user data that messes up the phone. A much more powerful reset is a deep reset.

Deep reset. Deep reset is not available on Series 40 phones, since user data and applications on those devices do not have direct access to the operating system, and hence, cannot crash the phone unless it is in bad shape—in which case, you should have it repaired.

If you need to delete all the contacts or images on a Series 40 phone, you have to do it manually. The Nokia PC Suite [Hack #15] provides a phone content browser that can help you speed up your cleaning procedures.

On a Series 60 device, a deep reset is equivalent to reformatting the C: drive and wiping out all the user data in the phone's internal flash memory. The data on the MultiMediaCard (MMC) card, however, is not touched. You can reformat the MMC card via the Options → Format mem. card menu in the Extras → Memory menu. You can force a deep reset on a Series 60 device in two ways:

- If your phone can still boot up, you can enter the service code *#7730# in the idle screen.
- If your phone does not boot into the idle screen, you need to hold the green Call key, the * key, and the number 3 key simultaneously while you power on the phone.

Although the MMC card should not be reformatted during a deep reset, you should probably take your MMC card out of the phone before doing a deep reset to be on the safe side.

You should see the word *formatting* on the screen during the deep reset.

It is very important that the phone has power while it is formatting. Do not take the battery out in the middle of formatting! In fact, I recommend that you connect the phone to an AC adapter before you do a deep reset.

Of course, after a deep reset "fixes" your phone's problems, you still need to restore some of the user data to make the phone useful again. That is more complex than simply resetting and requires you to plan a backup strategy before you start fiddling with the phone. Backing up and restoring give you the ability to roll the phone back in time to the stable and useable state before your latest failed experiments. In the rest of this hack, I will discuss data backup and restore strategies.

Automatic Backup and Restore

Nokia provides tools to automatically back up and restore the flash memory in your Series 60 phone (i.e., the C: drive). You can do it via a companion PC or via the phone's MMC card, which is not formatted during a deep reset. Automatic backup is easy to use and is safe in most occasions.

Use the PC Suite. The Nokia PC Suite [Hack #15] allows you to back up the entire content of the phone's internal memory (the C: drive) to a PC file. The full backup file includes the following:

- Contacts
- Calendar items
- Documents such as notes, email messages, and MMS messages
- Images and other media files in the phone's Gallery
- Settings including GPRS access points, email, and other connection settings
- Personal preferences such as wallpaper, ring tones, and speed-dial shortcuts
- Applications including Symbian C++ and Java programs

I recommend doing the backup periodically. If you mess up and have to do a deep reset, you can locate the closest clean backup file and restore it to the phone via the PC Suite. The PC Suite puts your backup files into *%SystemRoot%\Nokia\Phone Model\Backup* (e.g., *C:\Nokia\Nokia 3650\ Backup* for a typical backup for a Nokia 3650 phone).

The PC Suite also allows you to back up and restore a subset of data in the flash memory. You can choose any combination of the following four categories of data: contacts, calendar items, documents, and images.

Use the MMC card. You can back up everything in the phone's main memory to the MMC card via the Extras → Memory menu. Just choose the Options → Backup phone mem. menu and the data is backed up in a file, *E:\Backup\ Backup.arc*, on the MMC card (see Figure 4-4).

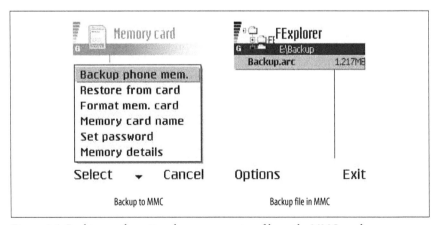

Figure 4-4. Backing up the entire phone memory to a file on the MMC card

When you decide to restore the phone memory from the backup file, just choose the Options → Restore from card menu in the Extras → Memory menu. The phone will reboot itself after the memory contents are restored.

The *Backup.arc* file is written over every time you perform a backup. So, by default, you can recover to the phone state only at the last backup time. To work around this, you can copy the *Backup.arc* file off the memory card after each backup and archive it on your computer periodically. If you need to recover to an earlier date, you can copy the correct *Backup.arc* file back to the memory card and then perform the restore operation.

Manual Backup and Restore

The automatic backup and restore solution is easy and effective. But it requires discipline to back up your data regularly and to remember to back it up before each experiment. It is also less useful if the backup interval is too long and too much user data is changed between backups. In addition, if your phone is infected with a virus before the last backup and you have just noticed the virus now, how do you know which past backup is the clean one that can be safely restored?

Using manual methods described in this section, you will be able to selectively back up some of your phone data right before the deep reset, and then restore it after the reset. Here, I assume that you can still boot up your phone and operate it.

> Even if your phone is fully functional and is not affected by malicious programs, you can still optimize the phone's memory by "refreshing" it. The manual backup process before refreshing gives you the opportunity to eliminate stale data and programs.

Connection settings. First, you should back up your current network settings, such as SMS server numbers, email settings [Hack #60], and GPRS access points [Hack #10]. You can simply go through the phone and write down those settings on a piece of paper for your records. After resetting, you can manually enter those settings back into the phone.

Some versions of the Nokia PC Suite (e.g., the PC Suite for Nokia 3650) provide a utility to back up and restore phone connection settings in a PC file. You can use the utility to automate the connection settings backup process.

PIM data. The phone's Personal Information Manager (PIM) data, such as contacts and calendar items, can be synchronized with PIM applications on desktop computers [Hack #36], [Hack #37], and [Hack #38]. You can synchronize them back to the phone after the reset.

If synchronization is not an option for you, you can use the backup and restore utility on the Nokia PC Suite, and request that it back up/restore the contacts and calendar data only (see the previous section in this hack).

Or, if you want to pick and choose which PIM items to back up, you can copy them to the PC via the Nokia PC Suite's phone browser. The PIM items appear as *.vcf* (for contacts) and *.vcs* (for calendar items) files on the PC. When you copy them back, the phone automatically converts them to the phone's internal format.

Documents. The best way to save documents—such as the email messages, MMS messages, audio/video files, and images—from being erased in a deep reset is to save them in the MMC card, which you can take out of the device when you do the deep reset. You can configure the Messaging and Camera applications to automatically save files to the MMC card. Please see "Manage Your Phone's Memory" [Hack #21] for more information on this subject.

If you do not save the phone documents in the MMC card, you can still copy and save everything in the phone's *Gallery* and *Messages* folders to a PC via the PC Suite before the deep reset.

Applications. If you have installed your applications on the MMC card, they will still be available on the phone's Main menu after the deep reset. However, the Application Manager program might not keep track of them anymore. So, you might lose the ability to update or remove those applications. For the Application Manager to work, you need to copy the contents of the *C:\System\Install* directory to the *E:\System\Install* directory on the MMC card. Those directories contain the installers (i.e., the *.sis* archive file) of applications installed on the device (see Figure 4-5). Since those system directories are not available in the Nokia PC Suite's phone browser, you have to copy them manually using a program such as FExplorer.

If you did not install applications on the MMC card and you have not backed up the entire phone memory using an automatic backup method covered in the previous section, you must reinstall all the applications from scratch.

HACK #24   Deal with Lost or Stolen Phones

Protect your wallet and privacy by preventing unauthorized access, even when you do not physically possess your phone.

It is easy to lose a mobile phone. Most of us have had the experience of forgetting the phone on our desk or in our car. Our kid might take it to school, or our spouse might take it on a trip. Sometimes, the phone might even get

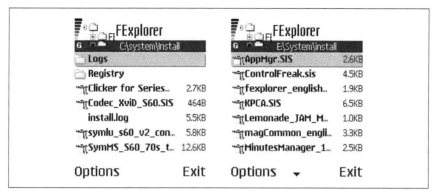

Figure 4-5. The application installation files required by the Application Manager program

stolen. A lost phone can expose you to big security risks, even if it is out of your physical control for only a couple of minutes:

- Someone might be able to obtain the sensitive personal and business information stored on the device.

- Someone might use the phone to make expensive calls or send messages to premier SMS service numbers and have the cost billed to you.

In this hack, you will learn how to protect yourself against information theft and unauthorized service access when you lose your phone.

 Remember to remove all locks, security codes, and personal data from your phone before you sell it on eBay! The best way to do this is to perform a hard reset and then format the memory card.

Lock the Phone

You can lock most Nokia phones (this is much more restrictive than locking the keys) by tapping the Power button, choosing "Lock phone" from the menu, and entering the lock code (see Figure 4-6). Once the phone is locked, you can perform only two actions:

- You can tap the left soft key, which is labeled Unlock on the screen, and enter the lock code again to unlock it. If you are in the dark, tap the Power button to turn on the LCD backlight.

- If there is an incoming call, the locked phone will ring and you can take the call by pressing the green Call button.

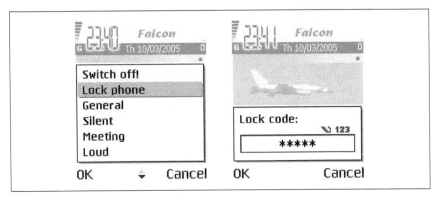

Figure 4-6. Locking a Nokia Series 60 phone

 If you just want to lock your keypad to prevent accidental dialing (or launching applications) when the phone is in your pocket, you can press the left soft-key and the * key at the same time. To unlock it, just press the left soft-key (or middle soft-key on devices with three soft-keys) and the * key again. No password is needed. Accidental dialing is especially a problem for phones that have the navigation pad sticking out like a joystick (e.g., the Nokia 6600).

You cannot dial any phone number or execute any application when the phone is locked. The default lock code in a factory-fresh phone is 12345. You can change the lock code in the Settings → Security → Phone and SIM → Lock code menu on a Series 60 device (see Figure 4-7) or the Settings → Security settings → Access codes → Change security code menu on a Series 40 device (see Figure 4-8). On earlier Nokia Series 60 devices, you might find those settings in the Tools program rather than the Settings program.

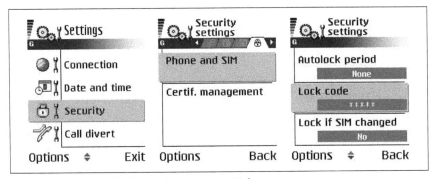

Figure 4-7. Changing the lock code on a Series 60 device

Using the security settings, you can also configure the phone to automatically lock itself after a period of time (see the "Autolock period" option in Figure 4-7).

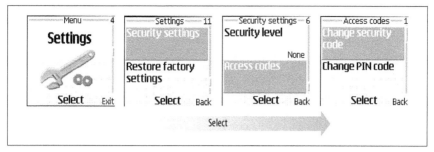

Figure 4-8. Changing the lock code on a Series 40 device

Protect the Memory Card

The memory card on a Nokia device can store large amounts of personal data, such as pictures and other media files. Even with a locked phone, someone can still physically remove the memory card and potentially read its contents using a card reader. You can use a password to protect the memory card so that it cannot be used in other devices or card readers.

On a Series 60 device, you can lock the memory card via the Options → Set password menu in the Extras → Memory menu (see Figure 4-9). On a Series 40 device, you can set the memory card password via the Options → Set password menu when you select the memory card icon in the Gallery menu (see Figure 4-10).

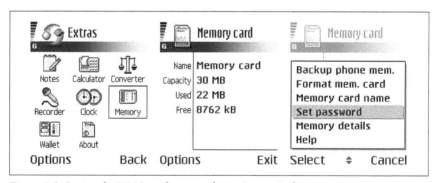

Figure 4-9. Setting the MMC card password on a Series 60 phone

Encrypt Sensitive Data

Many Nokia Series 40 and Series 60 phones include the Nokia Wallet application. On my Nokia 6600, the Wallet application is accessible via the Extras → Wallet menu. The first time you use Wallet, you will be asked to create a *wallet code*, which is used to generate an encryption key. Then you can store small bits of data, such as web site passwords and bank accounts,

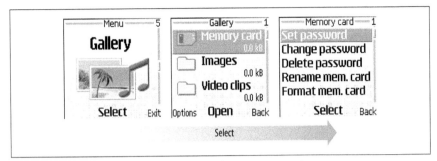

Figure 4-10. Setting the MMC card password on a Series 40 phone

in the wallet. The data stored in the Wallet application is encrypted and is accessible only to the person who knows the wallet code. On some recent phone models, the Wallet application can also integrate with the Services browser to save information you fill out on web forms.

 If you key in an incorrect wallet code three times in succession, the Wallet application is blocked for five minutes. The next three incorrect entries of the code will double the blocking time.

If you need to encrypt entire datafiles on a Nokia Series 60 phone, third-party applications such as SmartCrypto from SymbianWare (*http://www.symbianware.com/product.php?id=scrypto60&pl=n6600*) can help you. It encrypts any individual file on the Series 60 filesystem. Once the file is encrypted, you have to know the password to read its contents.

Restrict Calls

To protect against unauthorized calls, you can restrict the type of calls the phone can make or receive, via the Settings → Call barring menu on a Series 60 device and the Settings → Security settings → Call barring service menu on a Series 40 device (see Figure 4-11). For example, you can bar the phone from making expensive international calls when the device is roaming.

When you activate or remove those settings, the phone communicates with the network to update settings with the network operator (see Figure 4-12). You need to have the four-digit SIM PIN code to activate call restriction. Call your operator's phone support and ask for the Extended Services SIM password, if you do not have it. Once you have the SIM card PIN, you can change it from the same menus you use to change the phone's lock code.

An even more extreme measure is to restrict the phone to dial only a predetermined list of phone numbers. You can use the Settings → Security →

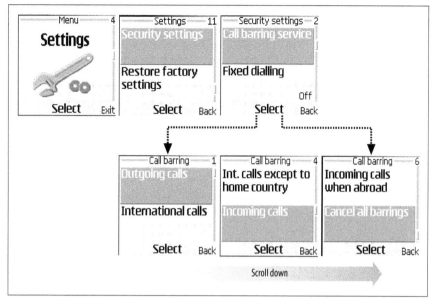

Figure 4-11. Restricting calls on a Series 40 phone

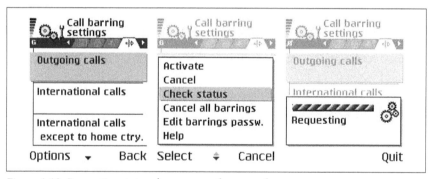

Figure 4-12. Requesting status changes over the network

Phone and SIM → Fix dialing menu on a Series 60 phone or the Settings →
Security settings → Fixed dialing menu on a Series 40 phone to do that. You
will be asked to select the allowed numbers from your Contacts list, and to
enter the SIM password to activate the setting.

Report a Lost or Stolen Phone

If your phone does get stolen, the first thing you should do is call the wireless operator and deactivate the SIM card so that the thief cannot make calls
against your account. You can also ask the operator to put the International
Mobile Equipment Identity (IMEI) number [Hack #5] of your phone into a

blacklist database. The blacklist database is a cooperative effort by wireless operators to bar blacklisted phones from making calls on any of the participating operators' networks. This greatly reduces the value of a stolen phone. However, the current blacklist implementations have some problems:

- In most countries, the blacklist is a national or regional database. The thief might be able to export the phone to other countries where it has not been reported as stolen.

- Wireless operators in the U.S. have not joined any blacklists. This has made the U.S. a haven for stolen phones.

In addition, an experienced mobile phone hacker can alter the phone's IMEI number via data cables. It is an illegal operation, but it is possible. Once the IMEI number is altered, the phone is no longer on the blacklist.

> If you live in the UK, you can register your phone with the National Mobile Phone Register via the web site *http://www.immobilise.com/*. This allows you to quickly report a stolen phone to the UK national blacklist and get notified when the police recover your phone.

Prevent SIM Card Changes

To prevent your stolen phone from being used by thieves, you can lock the SIM card to the phone. After you report the stolen phone and deactivate the SIM card via your operator, the thieves will not be able to insert a new SIM card and continue using the phone.

On a Series 60 phone, you can do that via the Settings → Security → Phone and SIM → Lock if SIM changed menu. Again, you will need to enter the SIM password obtained from the operator.

> No solution is completely secure. An experienced hacker can still reflash your phone via a data cable **[Hack #7]** and get rid of the SIM lock. But the SIM lock does make it more difficult for thieves to resell the stolen phone.

Protect Your Remote Data
Protect mobile phone data on the phone's companion PC and web accounts.

Mobile phone data is stored not only on the phone, but also on the phone's companion PC and web accounts. For instance, the Nokia PC Suite can copy everything in the phone's memory to a PC for offline processing or backup purposes. Voice mail messages are typically stored in the wireless operator's voice server. Mobile web portal sites also store a lot of sensitive data from the contacts or photo albums in the Camera application. Any comprehensive mobile security solution needs to protect this remote data as well.

Secure the Companion PC

Generally, you should not use a public PC in a library or web cafe to copy data from your phone. In theory, you can delete everything off the public PC after you are done and you won't leave a trace. But in reality, you can miss or forget to delete important data. The Nokia PC Suite can also store important information and settings (e.g., Bluetooth settings) in directories that are not obvious to most users. Hence, to avoid these potential information leaks, I recommend avoiding public PCs altogether.

You should also safeguard your PC from viruses and spyware, which can be used to steal data from your phone. A piece of good PC antivirus software, such as Norton AntiVirus from Symantec, will protect your Nokia PC Suite databases and synchronized Outlook databases from attacks.

Use Strong Passwords for Web Accounts

You should use strong passwords to protect online accounts for your mobile phone, especially the wireless operator's web account for your service, and web sites you frequently use to share photos.

Many web services also allow users to retrieve forgotten passwords by answering a personal question (i.e., the "password question"). It is important to use a password question that is not easy to guess. And don't forget: the answer to the question "what is your favorite pet's name" does not have to really be your favorite pet's name. It just has to be something you can remember.

In early 2005, it is believed that a cracker either guessed socialite Paris Hilton's T-Mobile account password, or exploited an SQL injection vulnerability on the T-Mobile portal web site [Hack #51] and reset her password. Regardless of how the attacker did it, Paris's account was compromised. After logging into her account, the cracker downloaded and then posted all her personal information, including her Contacts list and phone camera photos, on the Internet.

Be Aware of Caller ID Spoofing

The wireless operator's caller ID service embeds your phone number in every phone call and every network connection you make from your phone. A common assumption is that caller ID is always accurate, since it is controlled by the wireless operator. In fact, some services even use caller ID as their primary authentication mechanism. For instance, by default, T-Mobile's voice mail allows you to check messages without a password if you are calling from your T-Mobile mobile phone. Some web sites automatically sign into your account if you make an HTTP request from a phone number registered to your account. In many cases, your caller ID has become part of your identity.

However, in reality, caller ID is not completely secure. An experienced phone hacker can manipulate the network and change the caller ID to any number she wants—this is called *caller ID spoofing*. Some callback services, such as Star38 (*http://www.star38.com/*) and Camophone (*http://www.camophone.com/*), allow anyone to spoof caller ID for as little as five cents per call.

You should call your service providers and request to set passwords for all mobile phone–related network services. T-Mobile now allows customers to set an optional password on their voice mail account.

> Remember that you cannot always trust the caller's identity based on the caller ID. For instance, if you get a call with the caller ID indicating "Bank of America," it does not necessarily mean that the call is indeed from a Bank of America representative. You certainly should not disclose your bank account information based on the caller ID. In this case, you should offer to call the representative back using a published number for Bank of America.

Make and Receive Voice Calls

Hacks 26–32

For most users today, making and receiving voice phone calls is still the primary function of mobile phones. Nokia phones are equipped with special features to improve and enhance the voice phone call experience. In this chapter, I'll explore some voice-related hacks and special services.

HACK #26 Fast Access to the Dialer

Dial phone numbers quickly.

In today's world of electronic gadgets, we no longer have to remember phone numbers. Instead, we use electronic address books to locate contacts. The tight integration between the address book and the phone is actually one of the major selling points for mobile phones. But compared with dialing from memory, navigating the menu-based address book (a.k.a. the Contacts list on Nokia phones) is often a slow process. In this hack, I'll discuss some tips for making phone calls faster while you're on the move. If you're one of the many people who still dial by the numbers, you should be cured of that by the end of this hack!

Both Series 40 and Series 60 phones offer a quick shortcut from the phone's idle screen to the Contacts application. The Contacts application is typically mapped to the left soft key on a factory-fresh phone. If you customize the left soft key to another application, you can still go directly to the Contacts application by clicking the center key on the navigation pad (i.e., the joystick) on Series 60 devices, or the down key on Series 40 devices.

Speed Dial

Nokia mobile phones have built-in speed-dial support. Speed dial allows you to assign a phone number to any of the numeric keys on the phone's keypad. When the phone is displaying the idle screen, you can simply press a key and then the green Call button to make a call to the key's assigned speed-dial number. On Nokia Series 40 phones, you can also press and hold a key on the keypad and the phone will automatically call its corresponding speed-dial number.

> On some devices, speed dialing is also known as *1-touch dialing*.

To assign a phone number to a speed-dial key on a Nokia phone, first you need to select a contact from the Contacts list. Then, open the contact, highlight a phone number, and use the Options → Speed dial menu to assign that phone number to a key on the keypad (on some Series 60 devices, use Options → Assign 1-touch no). The 1 key is always assigned to the voice mail number from your wireless operator (it is stored in the SIM card) and you cannot change it. Figure 5-1 demonstrates the process on a Nokia Series 40 phone.

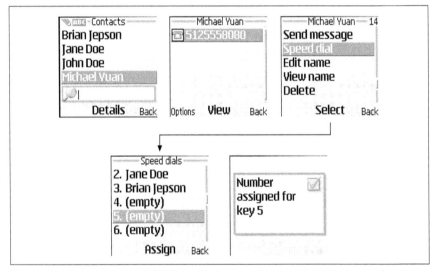

Figure 5-1. Assigning a speed dial from the Contacts list on a Nokia Series 40 phone

> If you try to speed-dial a key that is not yet assigned a number, the phone will prompt you to enter a number for that key.

On some Nokia Series 60 phone models (e.g., my Nokia 6600), you can also bring up the key map for the speed-dial assignments via the Tools → Speed dial menu (see Figure 5-2). If a key is assigned to a contact, the contact's thumbnail picture or name is displayed. For unassigned keys, the key number is displayed. You can use the Options menu to assign, remove, and change the phone number for each key.

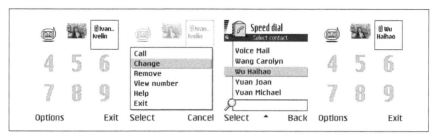

Figure 5-2. Setting up speed dialing on a Nokia 6600 device

The phone's built-in speed-dial application is easy to use but it holds only 10 phone numbers. If you need to speed-dial more than 10 numbers, you can use the speed-dialing feature in the SIM card directory.

Enhanced Speed Dial with the SIM Card

Your mobile phone's SIM card can store numbers and associated names just like a fancy landline phone's speed dial. The SIM card can support speed dialing for up to 250 numbers.

To start using this hack on a Series 60 phone, visit the SIM directory application usually found in the Tools → SIM directory menu. To add a number to the speed-dial list, select Options → New SIM contact, enter a name and number, and select the Done soft key.

To speed-dial numbers from your SIM card at any time, simply type in the location of the number (i.e., where it appears on the SIM's phone number list) and press the # key. For instance, to dial position 39, just dial 39# from the idle screen. Press the green Call button, and you're dialing!

If you tend to rotate among a few handsets frequently, you'll probably find this hack quite useful. Just remember that each SIM position can hold only one number, so plan ahead for that.

> When the caller ID display of the phone matches an incoming number to a name, it first searches the SIM directory and then the phone's Contacts list. So, if a phone number is associated with "Big Uncle Bob" on the SIM card and "Bob Robertson" in the Contacts list, your caller ID would show "Big Uncle Bob" as the incoming call.

You can set up SIM dialing on Series 40 phones in a similar manner. Usually you'll find the capability as an option when you click the Details soft key while browsing your contacts.

Search the Contacts

On Series 40 phones, enter the Contacts application through the Main menu and then select the Find or Search item from the menu inside the application to bring up the search screen.

> If you use the navigation pad shortcut to enter the Contacts application on a Series 40 phone (i.e., if you press the "down" direction from the idle screen), you will see a list of contacts, but not the menu. You cannot get to the search screen from there. You have to enter the Contacts application from the Main menu to search.

Then you can search the contacts by spelling out the first name in the text field. The list of matching contacts is updated as you type. After typing the first few letters, you can scroll down to the name of whomever you want to call and select the Details menu item. You will get a list of that person's phone numbers, which you can scroll through and dial using the phone's dial button.

Searching for a number to dial using the Series 60 Contacts application is slightly easier. When you bring up the Contacts application in Series 60 phones, the search text field is at the bottom of the main screen, so you can just start typing. You can also search by spelling either the first or the last name. After selecting the appropriate contact, you are shown a screen with all the information about that contact, and you can scroll down to the appropriate number and press the Call button to initiate the call. Figure 5-3 shows the search functions in the Contacts application on both Nokia Series 40 and Series 60 devices.

> In the search text fields, only "multitap" typing (e.g., to type C, press the 2 key three times) is supported. There is no T9 or other predictive text input support [Hack #58].

Advanced Contacts Search

As you can see, the default contacts search on Nokia phones is very primitive. It can search only the contacts' names (only first names for Series 40 devices) and supports only the slowest text entry method. But if you have a Nokia Series 60 phone, some additional software programs will drastically improve your search efficiency. One such program is Smart Dialer from Moov Software (available for evaluation and purchase at *http://www. moovsoftware.com/*).

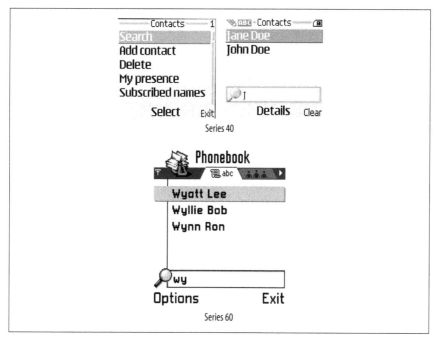

Figure 5-3. The search function in the Nokia default Contacts application for Series 40 and Series 60 devices

Once Smart Dialer is installed, it automatically runs in the background when you start up the phone. You can see it or stop it by holding the menu key and bringing up the application switcher.

With Smart Dialer running in the background, you simply start typing the number, first name, or last name you want to dial from the phone's idle screen. Smart Dialer automatically searches through your contacts and recently sent, received, and missed calls lists and returns a list of matches, narrowing the possibilities as you press more numbers. Smart Dialer's key feature is that it supports a very effective form of predictive text to match the search string to contacts. You can think of it as being similar to T9 text entry [Hack #58]. But instead of using a T9 dictionary to predict your input, Smart Dialer uses the names and phone numbers already in your Contacts list to come up with matches. For instance, you can type 642 to match all contacts in your Contacts list that have the following characteristics:

- Any contact with a combination of mno, ghi, or abc at the beginning of the first name or last name. For instance, Michael, Michele, and Nick all match this criterion.
- Any contact with a phone number that contains the numbers 642.

Figure 5-4 shows the Smart Dialer program at work. Notice that the dialed digits, 55, are matched to a number of contacts based on their first name, last name, and phone number patterns.

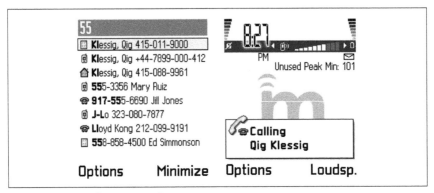

Figure 5-4. Using the Smart Dialer program

To make the search more efficient, Smart Dialer orders its search results by the number of times you've dialed the contact. You can also spell Rec (732) to get calls in your recently dialed, received, and missed lists.

To dial a selected contact, you simply press the green Call button. If instead you want to view or edit information about the contact, press the center key on the navigation pad.

Voice Tags

Another way to dial a number from your Contacts list quickly is to use a voice tag. This way, you do not even need to dial any numbers. You can add a voice tag to any phone number in your phone's Contacts list via the Options → Add voice tag menu item. To use voice dialing, you can press and hold the right soft key until the phone prompts you to speak. If your voice matches any of the recorded voice tags, the phone automatically dials the corresponding number.

> You can also use voice commands to launch applications on your smartphone [Hack #19].

The voice tag feature is very useful when you use your phone with a hands-free kit inside an automobile. You can start and end entire conversations without being distracted by the keypad and phone screen.

Use International Formatting for Numbers

We've discussed how to access the phone numbers in your Contacts list quickly. But those tips are useful only when you can use the phone numbers stored in the Contacts list. In particular, if you are a traveler, a number in your Contacts list might need a prefix when you call from another city or country. If that is the case, you cannot use any of the Contacts list features on your phone—a huge bummer, and a huge loss of productivity. One of the most important tips to speed up your dialing is to store properly formatted phone numbers, which are accepted in all areas you travel to, in the Contacts list in the first place.

Most wireless network operators allow you to use shortened phone numbers. In the U.S., you can typically just dial a seven-digit number to reach a local phone number. However, it is a bad idea to put seven-digit numbers in your Contacts list, especially if you travel a lot. For instance, the number 5551212 reaches different people in New York City and in San Francisco. At the very least, you need the 10-digit number, which includes the 3-digit area code, to dial anywhere in the U.S.

The situation is even more confusing if you travel to other countries. Ten-digit U.S. numbers will not work in the UK or in Germany. The best way to store phone numbers in your Contacts list is to use the international phone number format. To conform to the international phone number format, you simply need to add a +1 in front of a 10-digit U.S. phone number. Therefore, the phone number 5555551212 would be +15555551212. It doesn't route any differently if you call or send a Short Message Service (SMS) message to a number in that format from your home calling area—the operator knows what to do with it.

—Kamil Kapadia and Emory Lundberg

HACK
#27

Put a Face or Tune to the Caller

Use thumbnail pictures and ring tones to differentiate incoming callers.

Nokia mobile phones support the use of custom pictures or distinct ring tones to distinguish important callers or groups of callers. Using those features, you can give your phone a very personal touch. For instance, you might assign a special picture and ring tone for your mom. But more importantly, you can quickly tell who is calling without reading the caller ID. It is a lot faster to recognize a picture or a tone than it is to read words.

Use Contact Thumbnails

On most Series 40 and Series 60 phones, especially camera phones, you can assign a picture to each contact. Even if a contact has multiple phone numbers, it can still have only one thumbnail image. When a call is received from any phone number belonging to this contact, the picture is displayed with the caller ID information, including the contact's name, on the phone's idle screen.

To assign a thumbnail on a Series 40 phone, you must go to the contact's Details view. If the contact does not already have a thumbnail image, you can select the Options → Add image menu and then choose any image from the device's Gallery (see Figure 5-5). The device automatically resizes the image to the thumbnail size.

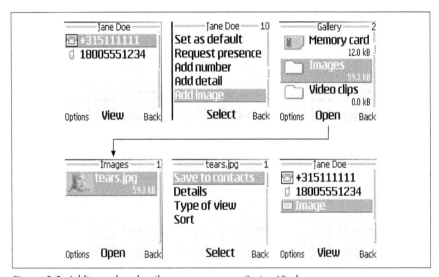

Figure 5-5. Adding a thumbnail to a contact on a Series 40 phone

If the contact already has a thumbnail image, you must first select that image and then use the Options menu to change or delete it (see Figure 5-6).

Figure 5-6. Handling existing thumbnails on a Series 40 phone

On a Series 60 device, you need to first enter the editing mode of a contact by selecting the contact and then the Options → Edit menu item. Then, use the Options → Add thumbnail menu to assign any image in the Gallery to this contact. If a thumbnail already exists for this contact, the new image will just overwrite the old one.

> The thumbnail picture feature can come in very handy when you meet someone at a conference. You can ask for his phone number and then take a picture of him together with the number. You will be able to put the name, the number, and the face together later, when you get home.

Use Ring Tones

On many Nokia devices, you can use ring tones to differentiate callers. You might assign a special ring tone to a person or a group of people. This way, you will be able to tell who is calling without checking the phone screen, so you can quickly decide whether to take the call.

Set a ring tone for each caller. On a Series 60 device, you can simply open the Contacts list (i.e., the address book) and open the contact person to whom you want to assign the ring tone. Then, from the Options → Ringing tone menu, you can select a ring tone for this contact (see Figure 5-7). Notice that all audio files in the Gallery appear on the list as well.

> Series 40 devices do not support per-caller ring tones. But they do support per-profile ring tones **[Hack #45]**.

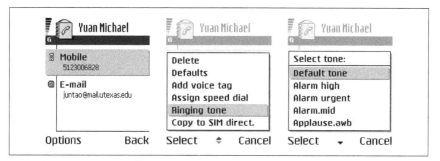

Figure 5-7. Setting a per-caller ring tone on a Series 60 phone

Set a ring tone for each caller group. On Nokia phones, the Contacts list (i.e., the address book) can organize contacts into caller groups **[Hack #43]**. Each contact can belong to multiple groups at the same time. For instance, on my phone, I have caller groups based on the contacts' relationship to me (e.g., the Family and Work groups) or the contacts' geographic location (e.g., the Austin and Dallas groups). You can specify a different ring tone for each caller group. Figure 5-8 shows how to set a ring tone for a caller group on a Nokia Series 60 phone.

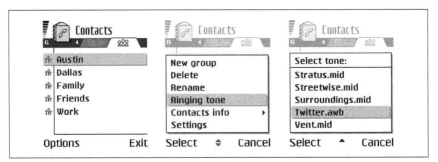

Figure 5-8. Setting a per-caller-group ring tone on a Series 60 phone

Series 40 devices also support per-group ring tones. To change the ring tone for a caller group on a Series 40 device, select the Contacts → Caller groups menu and then select a group (i.e., Family, VIP, Friends, Business, or any other you have named). Use the "Group ringing tone" option to select a standard ring tone, or use any third-party tone file in the Gallery.

Use Video Ring Tones

If you have a Series 60 phone, a Symbian application called Vision (available from *http://www.psiloc.com/index.html?id=168*) can bring your contact picture and ring tone to new levels. This program allows you to assign a video clip for each contact! When the contact calls, the video automatically plays on the screen, with full motion and sound. It is extremely cool and could really impress people around you.

The Vision program also supports large caller pictures. This is handy if you feel the default caller pictures are too small and are hard to see.

HACK #28 Use Prepaid Calling Cards

Prepaid calling cards help make your international travel easy and give you more control over account usage.

Subscription-based mobile phone services, while they are very convenient, are not suitable for everybody, or for every situation. Here are some example scenarios when you might not want to subscribe to a mobile phone service plan:

- You do not want to share your personal information and credit information with the operator.

- You do not want to be locked into a specific call plan or rate plan for an extended period of time.

- You travel abroad and do not want to pay for international roaming.

- You do not want to be surprised by the over-usage charges that are calculated only at the end of the billing cycle. For instance, you might want to put usage caps on the phones you give to your teenage kids.

Prepaid services are extremely popular in Asia and Europe. They are the best way to obtain mobile services when you travel to those countries. In the U.S., prepaid services are also becoming increasingly popular, as mobile phone services reach out to a broad population. Here is how to use a prepaid phone card to obtain services.

First, you need to purchase a valid SIM card from an operator that provides prepaid services in the area where you will be spending time. Put the SIM card into your unlocked phone **[Hack #7]** and turn it on to register on the network. The SIM card gives you a phone number but no services. The phone can make calls only to the operator or to emergency services at this point. The SIM card is relatively cheap (about $25 in the U.S.), and you can keep it for as long as you want. Using the same SIM card allows you to have a stable phone number.

Prepaid cards in foreign countries also give you a local mobile phone number for the locals to call.

Second, you need to purchase a prepaid card from the operator's retail store. The cards are also available in many local gas stations and grocery stores. The prepaid card has a face value of a number of available minutes. Once you've paid for the card, scratch the back of the card to reveal the secret PIN number.

Then, use the phone to make a call to the recharge number listed on the card. Listen to the instructions and enter the PIN when you are prompted to do so. The network operator will verify the PIN and unlock your SIM card for the number of minutes listed on the card.

Now, you can use the phone to make and receive phone calls until the prepaid minutes run out. At that time, you'll need to buy another prepaid card and repeat the process to recharge your SIM card.

Prepaid services are typically more expensive than subscription-based services in terms of per-minute charges. But if you do not use up all your plan minutes in a subscription plan, you might end up paying less using a prepaid service.

Use Calling Cards with Your Mobile Phone

Use a discount-rate calling card without the hassle of manually dialing PINs or even access numbers. Also, make the calling card work with your Contacts program!

Calling cards are good companions for mobile phones. Mobile phone calling plans, especially regional plans, sometimes have expensive long-distance rates. Even for national plans, the rates for international calls are often very expensive. Using calling cards, you can call in a local or a 1-800 access number, and then have the calling card service connect you to the destination number via a landline or even the Internet at very cheap rates. You'll still use up your plan minutes, but you won't pay as much in long-distance charges.

Long-distance calling cards discussed in this hack are different from prepaid mobile phone cards discussed in "Use Prepaid Calling Cards" [Hack #28].

However, making a calling card call on a mobile phone manually can be a real hassle. You have to key in a 10-digit access number; wait for the prompt; key in the PIN code (which is normally longer than 8 digits); wait for the prompt; and finally key in up to 16 digits of destination phone numbers (in the case of an international call). Try that when you are walking or driving! Fortunately, a couple of tricks are available that can make it easier to use calling cards.

Use Speed Dialing

With so many calling card services, you can shop around for the one that provides the best services and the best rate. Some calling card services provide two key features for easy mobile dialing:

No-PIN authentication
    Some cards allow you to register several preauthorized phone numbers with your account. If the calling card service caller ID detects that you are calling the access number from one of the authorized phone numbers, it automatically authenticates you without asking for your PIN number.

Speed dial for destination numbers

Some services allow you to assign two- or three-digit speed-dial codes for your frequently called numbers. Once you are authenticated and authorized to access your account, you can use those codes to dial long numbers quickly.

Using these two features, you can shorten a 36+ digit international call to 14 or so digits. They are great timesavers. We recommend you use calling services that provide those features.

Store Calling Card Numbers in the Contacts List

If you make frequent calling card calls, you can put the calling card access number, the PIN, and the destination number in one entry in your phone's Contacts list. Then, you can dial it quickly by locating the contact or even set it for speed dialing or voice commands.

But the problem is that the phone cannot just dial all the digits without a pause. The calling card service needs to accept the call after you dial the access number, before you can dial the PIN. It also needs to verify the PIN before you are prompted to dial the destination number. How do you put pauses between the access number and the PIN, and then between the PIN and the destination number, without human intervention?

Nokia phones have a nifty feature that puts pauses in a stream of digits. You just enter the phone number normally in the contact entry. When you need a pause between numbers, you click the * key three times until the letter "p" appears. If you need a longer pause, you can enter more than one "p" consecutively. Then you can go on to enter the next number after the pause. Figure 5-9 shows this process.

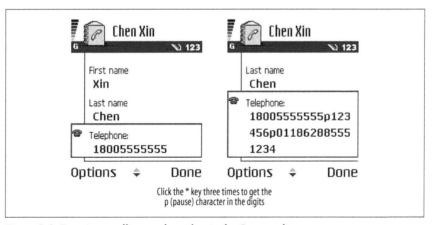

Figure 5-9. Entering a calling card number in the Contacts list

—Haihao Wu and Michael Yuan

Make Conference Calls

Your Nokia phone is a low-cost conference-call solution.

You can hold multiple conversations at the same time on your Nokia phone. That essentially gives you the capability to run conference calls right from your phone!

> The phone's idle screen displays each connected call in a text box. The forefront text box always shows the currently active call.

Receive Multiple Calls

If a call comes in while you are already in an active conversation, the phone beeps and flashes a text box for the incoming call. You can decide to take, reject, or ignore the incoming call, or simply put it on hold (see Figure 5-10).

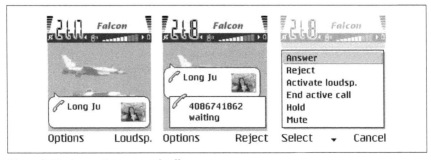

Figure 5-10. Answering a second call

If you do take the incoming call, the currently active call is automatically placed on hold. You can use the Swap soft key to switch between on-hold calls (see Figure 5-11).

Dial Multiple Calls

If you need to dial another number when you are in an active conversation, you can use the Options → New call menu. The phone will request that you enter the new phone number, and you can press the green Call button to connect the call (just as you would do for a regular call). Once the new call is connected, it becomes the currently active call, and all other calls are placed on hold (see Figure 5-12).

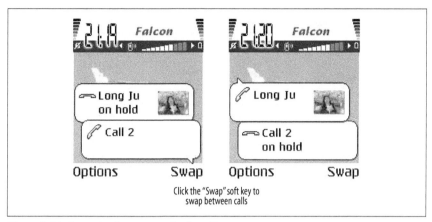

Figure 5-11. Swapping between multiple calls

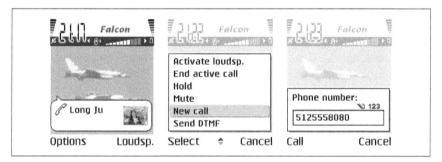

Figure 5-12. Making a new call while you are already on a call

Make Conference Calls

Once you have more than one connected call (i.e., several text boxes
appear on the idle screen), you can activate all of them at once by choosing Options → Conference (see Figure 5-13). This opens up a conference
and allows all callers connected to your phone to hear each other. Please
note that, for you to do this, your mobile operator and your service plan
must support conference calling. Also, when you look at your statement,
you'll see that the number of minutes is the actual number of minutes you
used, multiplied by the number of callers. Hence, it's a good idea to use
mobile-to-mobile minutes or night and weekend minutes for long conference calls.

> As a simple hack, you can also put the phone on loud-
> speaker via the Options → Activate loudsp menu item while
> the call is in progress. If the volume is loud and the room is
> quiet, all the people around the table will be able to hear the
> conversation and participate in the call.

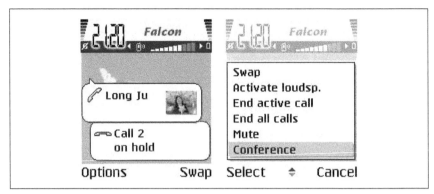

Figure 5-13. Starting a conference call

 ## Record a Phone Call
#31   Record important phone conversations on your Nokia smartphone.

From time to time, we often need to record phone conversations for business or personal reasons. For instance, when you order merchandise over
the phone, you probably want to keep a record of exactly what was agreed
on in case a dispute arises in the future. On a landline telephone, you can
use an answering machine to record phone conversations when you talk
with lawyers, accountants, technical support, or even sales agents. But for
mobile phones, the "answering machines" are located in the operator's
server room. They record messages for missed calls but do not provide a way
to record a conversation once you pick up your phone. How do you record a
phone conversation on your mobile phone?

 In some jurisdictions, the law requires you to notify the
other party that the conversation is being recorded.

On a Nokia Series 40 phone, you can record a phone conversation using the
Options → Record menu from the idle screen while the phone call is in
progress. The recorded conversation is saved as an audio clip in the phone's
Gallery. Most Series 40 phones support up to three minutes of recording
time.

Nokia Series 60 phones do not support phone conversation recording out of
the box. But third-party programs are available that add powerful recorder
functionality to Series 60 phones. In this hack, I discuss two such programs.

Extended Recorder

The Extended Recorder program from Psiloc (*http://www.psiloc.com/index.
html?id=167*) supports recording phone conversations on Nokia Series 60

phones. Unlike most answering machines on landline phones, the Extended Recorder program records into a digital audio file, which you can copy, share, and manage on a computer. The program supports several convenient ways to start and stop the recording. By default, you can use the Pen/ABC key to start and the green Call key to stop the recording at any time in the conversation.

> Extended Recorder provides more features than call recording. You can learn more about these features (and how to customize Extended Recorder's hotkeys) in "Record Audio" [Hack #75].

You can also configure the program to record all incoming calls from specific numbers in the Settings menu (see Figure 5-14). The Extended Recorder program needs to run in the background for the function to take effect. To put Extended Recorder in the background, just start it as you normally would and then press the red End key to return to the Main menu without exiting it.

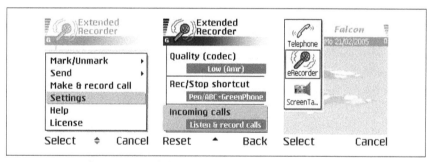

Figure 5-14. Configuring Extended Recorder to record incoming calls when it is running in the background

> To confirm that Extended Recorder is indeed running in the background, press and hold the "menu" key on the phone keypad. You should see the Extended Recorder icon in the icon list at the upper-right side of the phone screen. The last screenshot in Figure 5-14 shows this.

If you want to record an outgoing conversation, you can use the Make & record call menu to make the call (see Figure 5-15).

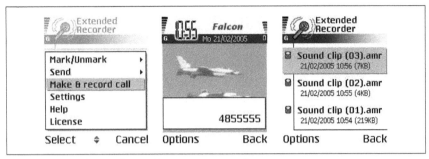

Figure 5-15. Making and recording an outgoing call

CallRecorder

CallRecorder (*http://www.symbianware.com/product.php?id=callrecorder60*) is another Symbian program for recording phone calls on Nokia Series 60 devices. Like Extended Recorder, it records both incoming and outgoing calls.

The unique feature of CallRecorder is that it provides finely grained control over exactly which calls are recorded. For instance, you can configure it to record all calls made to numbers with the 512 area code, or all calls received from a certain person in your Contacts list. You can also decide whether to record on a per-call basis. The CallRecorder program can prompt you to elect whether to record the call when the caller or called number matches a certain pattern.

> On the other end of the spectrum, SymbianWare's CallCheater program (*http://www.symbianware.com/product. php?id=callcheater60*) generates static sound and injects it into your conversation. That simulates bad signal reception and allows you to drop an unwanted call without offending the other party (unless they find this book on your coffee table and get wise to you).

Use the Mobile Phone As an Audio Tour Guide

HACK
#32   Get more out of your trips by listening to tour guides on your phone.

Wouldn't it be wonderful to let comedian and actor Jerry Stiller take you on a tour of New York's Lower East Side, the very neighborhood he grew up in? Wouldn't it be great to hear, when you are on the steps of the Lincoln Memorial, a recording of Martin Luther King Jr.'s "I Have a Dream" speech that he gave on that same spot?

In many museums, historic sites, and parks, you can have such experiences by participating in self-guided audio tours. But often you have to wait in a long line to rent an audio guide device first, and then you have to figure out how all the buttons function before you can actually go on the tour. For many visitors, this is just too much of a hassle. And for museums and parks, it costs too much to manage and maintain all those devices. Many smaller museums and parks simply do not have enough funds or personnel to provide such tours. Consequently, visitors' experiences become less informative and less engaging.

Fortunately, you can use your own mobile phone as an audio tour device. You don't need to rent a device, or try to figure out which buttons to push. Just dial a number provided by the places you visit, and you can listen to the audio tour right away.

 On a related note, when you are in a new city, you can usually call 411 (or some other operator-specific service number) from your mobile phone to get information about local attractions, driving directions, and additional local information.

This kind of mobile-phone-based tour is already available in many urban cities, national historic sites, zoos, museums, and national parks. They deliver audio content specially designed to enhance the visitor experience. (See Table 5-1, later in this hack, for a list of such tours.) For users, key benefits of mobile phone audio tours are:

Convenience
Your mobile phone is always with you. No device to rent, no guidebook to buy. And you don't have to learn how to use unfamiliar rented devices, or worry about loss and damage, so a mobile phone audio tour can reduce the anxiety you might feel when using rented or unfamiliar devices.

Flexibility
Since the service is voice based, it works with just about any phone and any mobile operator. Plus, the tour is available to you at any time. You can take the tour at your own pace and schedule. You don't need to wait for a tour guide.

Unique perspectives
Interesting things do not just happen in Times Square or on the Golden Gate Bridge. Because of the flexibility of a mobile phone audio tour, it can take you to places and tell you stories that are not normally available in museums or guidebooks. For instance, in the city of Toronto, you can access a unique cell phone audio tour for free. The tour

provides a rich collection of personal oral histories submitted by the city's residents. In the tour, you get to know the city's ordinary people and hear their remarkable stories, told in their own voices. This is certainly not available in usual guided tours in Toronto.

Vivid

Because of the audio format, the tour can be very vivid. Historic audio (such as an important speech), original music, and expert interviews can all be incorporated into the audio. For example, tourists walking by Ford's Theater can listen to a reenactment of President Lincoln's assassination. This is the experience you normally cannot have by just reading a guidebook.

Mobile phone audio tours can be easily programmed to provide multilingual content. They also can be very useful in large outdoor spaces, such as entire cities or large national parks that locally networked audio devices cannot cover.

A research institute, Touch Graphics, has developed a mobile phone–based tour to improve accessibility to museums and exhibits for blind and visually impaired visitors. The museum sets up a network of wireless audio beacons at key destinations in the exhibit space. Using their familiar mobile phones, users can call a toll-free number and select a personal *ping* sound. Then they choose a destination, and use their phone to trigger the personal ping sound from the beacons at the destination. By following their personal sound, they can navigate through the exhibit independently. When they reach their destination, the phone becomes an interactive audio guide, providing explanations about the exhibition.

Make It Work

Information about an audio tour is usually available from your destination's web site, banners, or brochures you can pick up at the visitor's center or entrance. You might also want to print out or obtain the audio tour map.

Now, dial the dedicated phone number specified on the tour information. Follow the voice instructions to select an audio tour track and fee plan. Sometimes, multilingual tours and special tours for children are available. You can choose accordingly through your phone. After that, input your payment information. Usually the fee is charged through the phone when you input your credit card number. You might also be able to put the charge on your mobile phone bill with the wireless operator.

When you explore the park or museum, use your mobile phone keypad to input the stop number corresponding to locations of interest. The stop numbers are usually located on the tour map and can also appear on signs around the site.

Tips and Tricks

Turning on your phone's speaker can enable multiple people to enjoy an audio tour, using only one mobile phone. This is fun for parents and kids to participate in the tour together.

Once you purchase your tour, you can access it within seven days. (Times might vary in different places. Some places allow three days.) Some tours allow you to listen twice.

After one stop, just hang up and call again when you get to the next stop. You can spend as much time as you want in between and also save minutes on your calling plan.

Note that the tour price tag does not include the charges to your cell phone plan and roaming fees, etc. So, check your cell phone plan and remaining minutes to know beforehand what the cost will be. With cheaper airtime minutes, especially with free minutes on weekends, cell phone audio tours should become more and more affordable.

Prices and Availability

Table 5-1 lists some of the currently available audio tours for mobile phones.

Table 5-1. Available mobile phone audio tours

City	Features	Phone number	Cost	Web site
New York	Tour of New York's Lower East Side (narrated by Jerry Stiller); Lower Manhattan (narrated by Sigourney Weaver)	212-262-8687	$5.99	http://www.talkingstreet.com/
Boston	Tour of Boston	617-262-8687	$5.99	http://www.talkingstreet.com/
Toronto	Personal Oral History storytelling	Various; see tour guide	Free	http://murmurtoronto.ca

Table 5-1. Available mobile phone audio tours (continued)

City	Features	Phone number	Cost	Web site
Sacramento Zoo	Mobile Phone Safari in the zoo	703-286-6545	$3.99	*http://www.saczoo.com*
Minute Man National Historical Park	The story of April 19, 1775, the first day of the American Revolution	703-286-2775	$5.99	*http://www.nps.gov/mima/*
San Antonio, Texas	Audio tour of the Alamo	703-286-6523	$6.99 (additional charge for a longer program)	*http://www.spatialadventures.com/*
Denver	Two-mile walking tour of lower downtown Denver	703-286-6365	$8	*http://www.rmaguides.com*

Mobile phone–based audio tours are not going to replace live guides and tour books. But they will become an increasingly popular approach for you to enjoy your visits to cities, parks, zoos, and museums. In the future, as mobile location technology and 3G wireless networks evolve, audio tours could even incorporate mobile phone–based maps and directions, videos, and interactive messaging, making the tour experience even more interesting and informative.

Another mobile phone tour guide service is Grafedia (*http://grafedia.net*). It provides community-contributed multimedia content to your phone via MMS messages, and it is free of charge. When you are at a new location and see graffiti with email addresses ending with "@grafedia.net," you can send an MMS message to that address **[Hack #59]** and receive related content (images, text descriptions, audio and video clips) via MMS. You can also upload your own content to the site and associate them with your own graffiti. See the site for detailed instructions.

—Ju Long

Exchange Data with Computers
Hacks 33–39

Smartphones are most useful when they work together with desktop or notebook computers. Computers have powerful software for storing and managing data, and mobile phones provide data access anytime, anywhere. This winning combination has been proven in the real world. But for the phone and the computer to work together, first they need to exchange data with each other.

In this chapter, I'll explore how to exchange and synchronize files, contacts, and other personal information items between Nokia mobile phones and desktop computers.

Exchange Files

You can exchange files between a Nokia mobile phone and a computer in a variety of ways, including via the Nokia PC Suite, Bluetooth file transfer, infrared beaming, email, MMS, and web downloading and uploading.

In this hack, I'll discuss ways to send and receive files to and from a mobile phone. Why do you want to do that? Well, here are some example scenarios:

- If you have a camera phone, you probably want to download your pictures to your PC or web site for editing or sharing **[Hack #67]**.

- You might want to customize your phone with wallpaper **[Hack #46]** and ring tones stored in media files **[Hack #45]**.

- You might want to enhance your phone with additional software **[Hack #16]** and **[Hack #17]**.

- You might want to exchange contacts (i.e., business cards) and calendar items between the computer and the phone via VCF-formatted files. The native Personal Information Manager (PIM) applications on the computer or on the phone can process these files once they arrive.

The list goes on. But this gives you an idea why file exchange with a mobile phone is important. As you can see, it is often the basis for further hacks.

Once a file is transferred to your phone, you can use the appropriate program to locate and process the file. For instance, media files are handled by the Gallery program, application installation files are handled by the Application Manager program, and VCF files (business cards) are handled by the Contacts program. If a file has an unknown format, you can locate and examine it using a file browser program such as FExplorer [Hack #20].

Use the Nokia PC Suite

The easiest way to exchange files between a Nokia phone and a Windows-based PC is to use the Nokia Phone Browser program in the Nokia PC Suite. The program adds a *Nokia Phone Browser* folder to the root directory in the Windows File Explorer (see Figure 6-1). Each device that is connected to the PC Suite shows up as a subdirectory in the *Nokia Phone Browser* folder. Each device directory is identified by its name (e.g., its Bluetooth name if the device is connected via Bluetooth).

You can drag files into and out of the mapped device folders in the Windows File Explorer. The changes are automatically synchronized to the device. In addition to the phone browser, the Nokia PC Suite also provides audio, image, and video manager programs so that you can manage mobile multimedia files on your phone.

> You can also use the Nokia PC Suite to transfer and install Java and Symbian application installation files into your phone [Hack #16] and [Hack #17].

The advantages of the Nokia PC Suite are its easy-to-use interface and its support for multiple connection methods, including Bluetooth, infrared data connection, USB cable, and serial cable. However, as I discussed in "Use the Nokia PC Suite" [Hack #15], the Nokia PC Suite does not fit all usage scenarios. In the rest of this hack, I will discuss other ways to exchange files between a Nokia phone and a computer.

Bluetooth

The Bluetooth File Transfer service allows any two Bluetooth-enabled devices to exchange files [Hack #11].

To send a file (e.g., an image file in the Gallery application) from your mobile phone, navigate to the file and choose Send → Via Bluetooth from the Options menu. Then, choose a device from a list of local devices to send

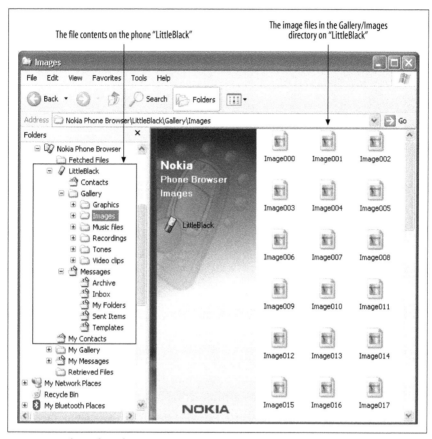

Figure 6-1. The Nokia Phone Browser program in the Nokia PC Suite

the file to. The devices are identified by their Bluetooth names. Figure 6-2 shows how to send a file from a Nokia phone to a computer.

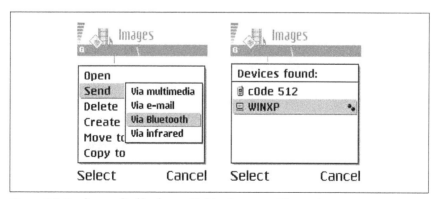

Figure 6-2. Sending media files from a Nokia phone over Bluetooth

When a Nokia phone receives an incoming file, the file is delivered to the Inbox. When the user opens the message, the phone automatically identifies the file type and processes it accordingly. For example, a received image is saved to the Gallery, a business card is stored to the Contacts list, and an application installation package is executed. To send a file *from* a computer, you need to use an operating system–specific Bluetooth file exchange utility, described in the following sections.

> You can send only one file at a time over Bluetooth. Hence, this method can be tedious if you need to transfer many files (e.g., photos from a multiday trip). The ZipMan application, available from Symbian (*http://www.wildpalm.co.uk/ZipMan7650.html*), allows you to create a zip archive of multiple files on your phone and send them all together to a computer. If you need to send many files from a PC to your phone, you can also zip them together first and then use Zip-Man to extract the files from the archive once the phone receives the zipped file.

Windows versions before XP Service Pack 2 (SP2). With the WIDCOMM Bluetooth utility installed [Hack #12], you can use the File Transfer service to send and receive files. You can right-click the Bluetooth icon in the system tray to bring up the pop-up menu that lists all available services. The received files and business cards are automatically stored into specified directories on the local disk. You can configure those destination directories in the Bluetooth configuration tool. Figure 6-3 shows the pop-up menu and the configuration window for the destination directory.

Windows XP SP2. On Windows XP SP2 [Hack #12], a Bluetooth File Transfer Wizard is located under the Start → Programs → Accessories → Communications menu. The wizard allows you to send and receive files to and from other Bluetooth devices. If you choose to send a file, the wizard first prompts you to choose a target device and then asks you to select the file to send from the PC. If you choose to receive a file, the PC listens for the incoming file and prompts you to save it once it arrives. Alternatively, you can right-click the Bluetooth icon in the system tray and select the "Send a file" or "Receive a file" menu item to directly invoke the appropriate functionalities in the Bluetooth File Transfer Wizard. Figure 6-4 shows the dialog boxes for sending a file and saving a received file.

> You have to click the "Receive a file" choice to prepare the PC to receive files over Bluetooth. Otherwise, the device will show a Connection Failed message when you try to send a file to the PC.

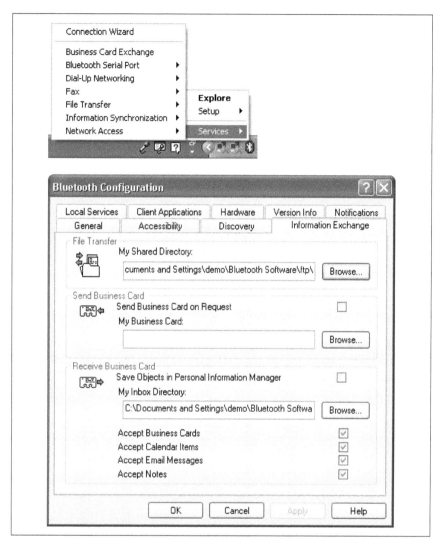

Figure 6-3. Sending and receiving files over WIDCOMM Bluetooth software

Mac OS X. Mac OS X has a Bluetooth File Exchange program under the */Applications/Utilities* directory. You can also launch it from the Bluetooth menu item (if you've enabled it in System Preferences → Bluetooth) by selecting Send File from the Bluetooth menu item. You can use this program to send files to devices. The Bluetooth settings management tool in System Preferences allows you to specify where to store the incoming files in the local computer. Figure 6-5 shows the file-sending window in the file exchange utility, and the configuration window for how to save the incoming files.

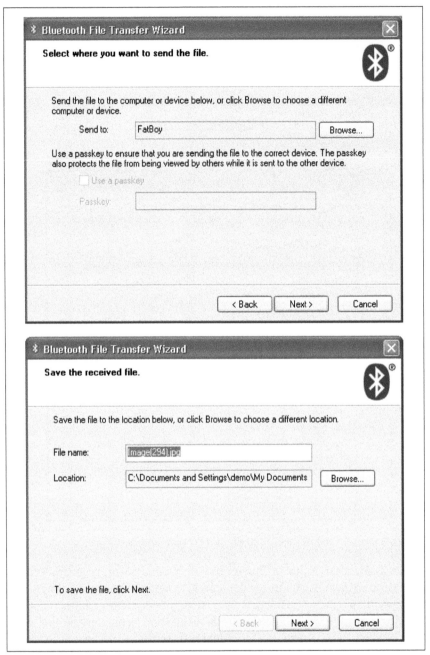

Figure 6-4. Sending and receiving files over the Windows XP SP2 Bluetooth File Transfer Wizard

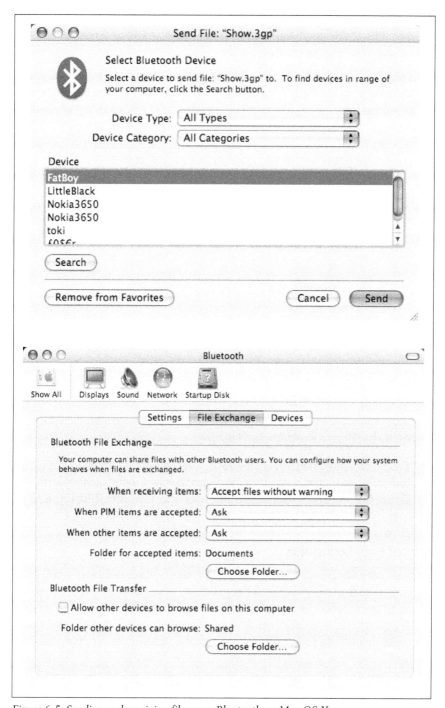

Figure 6-5. Sending and receiving files over Bluetooth on Mac OS X

Linux. If you run the KDE or GNOME desktop systems on Linux, you can use their Bluetooth graphical user interface (GUI) tools to send and receive files to and from mobile phones:

- You can download the KDE Bluetooth Framework from the project's web page at *http://kde-bluetooth.sourceforge.net/*. Figure 6-6 shows a Bluetooth incoming file received by the KDE Bluetooth Framework GUI.

Figure 6-6. Receiving a file over Bluetooth in KDE

- You can download the GNOME Bluetooth Subsystem from the project's web page at *http://usefulinc.com/software/gnome-bluetooth*. Figure 6-7 illustrates sending a file over the GNOME Bluetooth GUI tool.

Infrared Data Connection

Another popular choice is to "beam" files via the Infrared Data port. This method is particularly effective for transferring files between two IR-enabled Nokia phones. Since IR communication requires a clear line of sight, first you have to make sure the two devices' IR ports are facing each other and that there is nothing in between them. On the sending device, you can just choose the Send → Via infrared menu (refer back to Figure 6-2) for the target file; on the receiving device, the IR port must be turned on and the incoming file will be saved into the message Inbox, just as with Bluetooth file transfer.

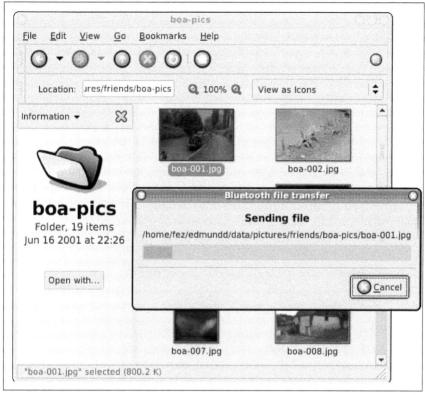

Figure 6-7. Sending a file over Bluetooth in GNOME

Using IR to transfer files between a phone and a computer is more complex, since most computers do not come with an IR port. You can buy an add-on IR adapter and plug it into the computer (usually via a USB port). The device driver that comes with the IR port should include a file transfer utility.

MMC Reader and Writer

Many newer Nokia phone models have expansion slots for MultiMedia-Card (MMC) cards, which provide additional memory space for images, multimedia files, third-party applications, etc. For some devices, such as the original N-Gage, the MMC card is directly accessible from a computer. When you hook an N-Gage to a PC via a USB cable, the MMC card appears as a removable disk in the Windows OS. For other devices, you can buy a generic MMC reader/writer and connect it to your computer via USB. In both cases, you can manipulate the files and directories via Windows Explorer, Mac OS X Finder, a Linux file manager, or the good old command line.

If you do connect in this way, you might find some interesting files (*.DS_Store* and *.Trashes* on a Mac, for example) on the MMC card. My advice is to ignore these, since they will be re-created each time you plug them in, and they do no harm.

Email

Email attachments are very popular for sending files across the Internet. You can use email attachments to transfer files to and from mobile clients as well. In "Send and Receive Email on Your Phone" **[Hack #60]**, I discuss how to set up and use your phone's native email client. The Nokia native email client allows you to attach media files from the device's Gallery to outgoing email, or save media file attachments in incoming messages to the Gallery. Using more advanced email client applications, you might be able to send and save any attachment file outside of the Gallery. Then you can manipulate those files with programs such as FExplorer **[Hack #20]**.

MMS

You can compose MMS messages on your Nokia phone and attach media files from the Gallery to them. You can also save file attachments in incoming MMS messages to the Gallery. MMS messages can transfer files from device to device, or from device to email address **[Hack #59]**. You need special software to send MMS messages from a computer to a phone. From the user's perspective, using MMS to transfer files is very similar to using email, except for the following differences:

- MMS messages are pushed to the target device. The user sees the message as it arrives. There is no need to check the Inbox from time to time.

- MMS is usually easier to configure than email. Most devices are sold with MMS preconfigured.

- Since MMS traffic is a low priority in wireless networks, it might take some time (perhaps tens of minutes, or even longer) to reach the destination device. In addition, the MMS message might not arrive at all, if it takes more than 24 hours to deliver it.

- If you're not on an unlimited data plan **[Hack #4]**, MMS traffic might be cheaper than email traffic over the General Packet Radio Service (GPRS) network.

- MMS messages are typically limited to a maximum of 100 KB.

Web Download and Upload

The last approach is to use a personal web site to share content between a PC and your phone. For instance, you can upload a file from a PC to the site and then use your phone browser to download it to the phone. The added benefit to this approach is that it allows other people to share files with you. For more details, please see "Create a Mobile Web Site" [Hack #53].

HACK #34 Transfer Contacts from a Computer

Creating business cards on a desktop computer and then sending them to your phone helps you avoid lots of typing on the cramped mobile phone keypad.

The Contacts (a.k.a. Address Book or Phone Book) program is one of the earliest and most successful applications on Nokia phones. Managing contact phone numbers right on the phone is clearly a convenience most people appreciate. In early mobile phones, the on-device storage space was extremely limited. So, the "contacts" were limited to pairs of names and phone numbers. As mobile phone technology evolved, modern Nokia Series 40 and Series 60 smartphones offered a much more advanced address book. The following list is just a sample of notable features in these advanced address books:

- You can store multiple phone numbers, email addresses, instant messaging IDs, and physical street addresses for each person.
- Each contact entry can hold a picture (e.g., a head shot or an icon) of the person.
- For each incoming call, the phone matches the caller ID with contacts in its Contacts list, and displays the caller's name and picture if a match is found [Hack #27].
- Contacts can be organized into groups and assigned distinctive alert tones [Hack #27].
- You can search for contacts based on the contacts' details [Hack #26].

Each contact entry in your phone's Contacts list is also known as an electronic business card, which now holds more information than a real paper business card. However, despite the evolution of the phone's Contacts list, the mobile keypad has changed little over the years. It is still slow and error prone in terms of text entry—indeed, it is a major pain to enter all that text-based information into a contact on the phone.

Your desktop or notebook computer comes to the rescue here. You can create contacts on a PC using a full-size keyboard and then send them to your phone. Or, if you already have a contact on your phone, you can easily send it over to other people's devices and save them from having to type it in by hand. In this hack, you will learn all about business card exchange with Nokia phones. I do not discuss contacts synchronization in this hack. Synchronization is covered in several separate hacks later in this chapter.

Use the Nokia PC Suite

The Nokia PC Suite includes a Contacts Editor program. The program's main window is a contact entry form. To create a new contact, just fill in the information in the blank fields. You can leave any of the fields blank. You can also add a thumbnail picture and notes to the contact (see Figure 6-8).

Figure 6-8. Entering a new contact in the Nokia Contacts Editor

After you are done, click the Save button and save the contact to the *Contacts* directory on a connected phone via the phone browser [Hack #15]. You will be asked to synchronize the new contact with the phone (see Figure 6-9).

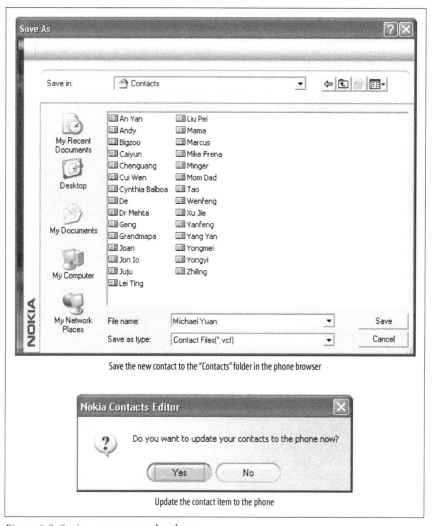

Save the new contact to the "Contacts" folder in the phone browser

Update the contact item to the phone

Figure 6-9. Saving a contact to the phone

The Nokia Contacts Editor program can also handle very complex business cards with multiple addresses for each contact. Click the Advanced button in the entry form to switch to that view. Again, you are free to leave any of the fields in the form blank.

Using the Nokia phone browser, you can double-click any contact item in the phone's *Contacts* directory to edit it and save it back to the phone (see Figure 6-10).

Figure 6-10. Editing existing contacts on the phone

Remote S60 is another Windows program that allows you to control a Series 60 phone directly from a PC over Bluetooth, IR, or data cables. You can use it to update your phone's Contacts list from a PC. But Remote S60 does much more than that—it lets you use your PC keyboard to take complete control over your Series 60 phone, and even displays a copy of the phone's screen on your PC. So, you can use it anytime you need a break from typing on the small keyboard. You can download and purchase the Remote S60 software from *http://mobileways.de/M/1/3/0/*.

Use the Mac Address Book

The Mac OS X default Address Book program is automatically Bluetooth aware. If Bluetooth hardware is installed on your Mac, a Bluetooth icon is visible in the OS X Address Book program's main window (see Figure 6-11). You can click the Bluetooth icon to connect the Address Book application to your paired phone (specifically, the one you chose to use with the Address Book when you paired). If no phone is paired, you can select one of the Bluetooth phones in range. Such pairing allows the Mac computer to dial phone numbers, receive and redirect calls, and send and receive Short Message Service (SMS) messages via the phone [Hack #63].

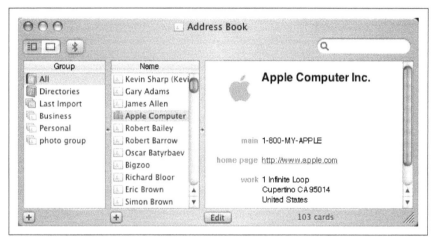

Figure 6-11. The Mac Address Book application with the Bluetooth button

To send a contact in the Address Book to a phone as a business card, first choose the contact, select the Card → Send This Card menu item, and then choose the recipient device from the Bluetooth device list (see Figure 6-12).

 The Card → Send This Card option is not available if Bluetooth hardware is not installed on the Mac.

On the phone, when you open the message, you'll see its contents. Choose Options → Save business card to save it to the Contacts list (see Figure 6-13). Please note that when you use Mac Bluetooth to send business cards, the thumbnail image and the Note field in the Mac Address Book are lost.

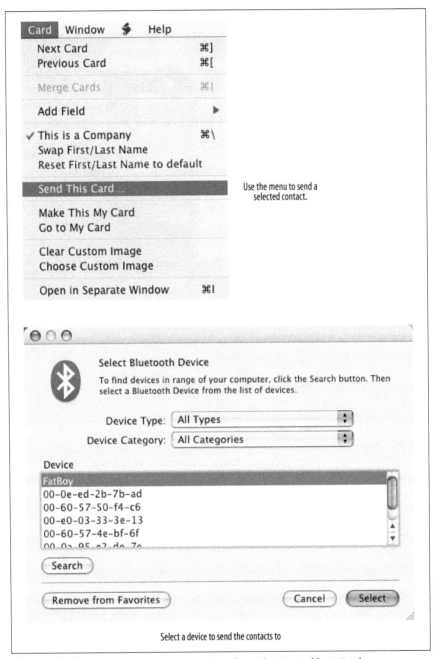

Use the menu to send a
selected contact.

Select a device to send the contacts to

Figure 6-12. Sending a contact item to a phone from the Mac Address Book

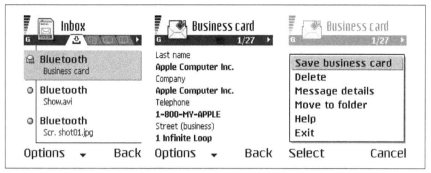

Figure 6-13. Saving a business card on the phone

Use gnokii on Linux

gnokii is a cross-platform open source interface to Nokia devices. It runs a variety of hardware and operating system platforms, including Linux and BSD. gnokii features a sophisticated command-line interface for exchanging contacts, calendar items, and even SMS messages between a PC and a connected Nokia device. You can download the gnokii software and read its documentation at *http://www.gnokii.org/*.

> A drawback of gnokii is that it does not integrate with the existing address book program on the host computer.

Several GUI wrapper programs have been developed to make gnokii's command-line interface user friendly. You can see their screenshots and find the download links at *http://www.gnokii.org/screenshots.shtml*.

Send Contacts as Files

So far, I discussed Windows-, Mac OS X-, and Linux-specific solutions for sending business cards to a Nokia phone. A generic and platform-independent method is to directly transfer the datafiles that represent business cards.

The ISO defines a standard file format for representing business cards. On most computers, the file has a filename suffix of *.vcf*. Most address book applications on Windows, Linux, and Mac OS X support importing and exporting contacts from and to *.vcf* files. In fact, the Nokia Contact Editor in the PC Suite saves and edits *.vcf* files in the phone browser's *Contacts* folder.

A Nokia mobile phone can open *.vcf* files in incoming messages (email, Bluetooth, or MMS) and save them to the Contacts list (refer back to

Figure 6-9 for an example). It can also export any Contacts item to a *.vcf* file and send it to another device.

The thumbnail image, if there is any, can be included in the *.vcf* file. Please note that not all address book applications generate Nokia-compatible *.vcf* files with thumbnail images. So, if your phone complains about an "unsupported format" when trying to open an incoming *.vcf* file, you should consider transferring a text-only business card.

Use a Web-Based Mobile Portal

Some wireless operators' portal web sites support remote provisioning of contact data. You can simply load a web form in your desktop computer browser to fill out the contact details and then click the "send" button (see Figure 6-14). The contact information is sent to your phone number via a special SMS message. You can review its contents in the message Inbox and then save it to the Contacts list on the device (see Figure 6-13 for an example).

The benefit of the web-based approach is that it is platform independent and requires no configuration. You can use any computer browser to access the portal and send the SMS message to any phone. However, most web portals do not support thumbnail images in business cards. Also, depending on your service plan, the wireless operator might charge a small fee (e.g., 10 cents) for each SMS message received on your phone. So, use this option with caution if you need to send a lot of business cards.

Share Contacts Between Phones

Exchanging electronic business cards in meetings is not only cool, but also efficient and reliable.

Often, we need to send contacts from one phone to another to avoid retyping the information. For instance, if you meet someone in a conference hall, instead of typing his address on your phone keypad, you can ask him to send his electronic business card directly to your phone. Then, when you go home, you just send the card to your PC for backup.

Exchanging business cards via Bluetooth is great if you forget to bring a business card or run out of them at a meeting.

Send Contacts from a Nokia Phone

In your phone's Contacts application, just highlight or open the contact you want to send, and then select Options → Send... on a Series 60 device or

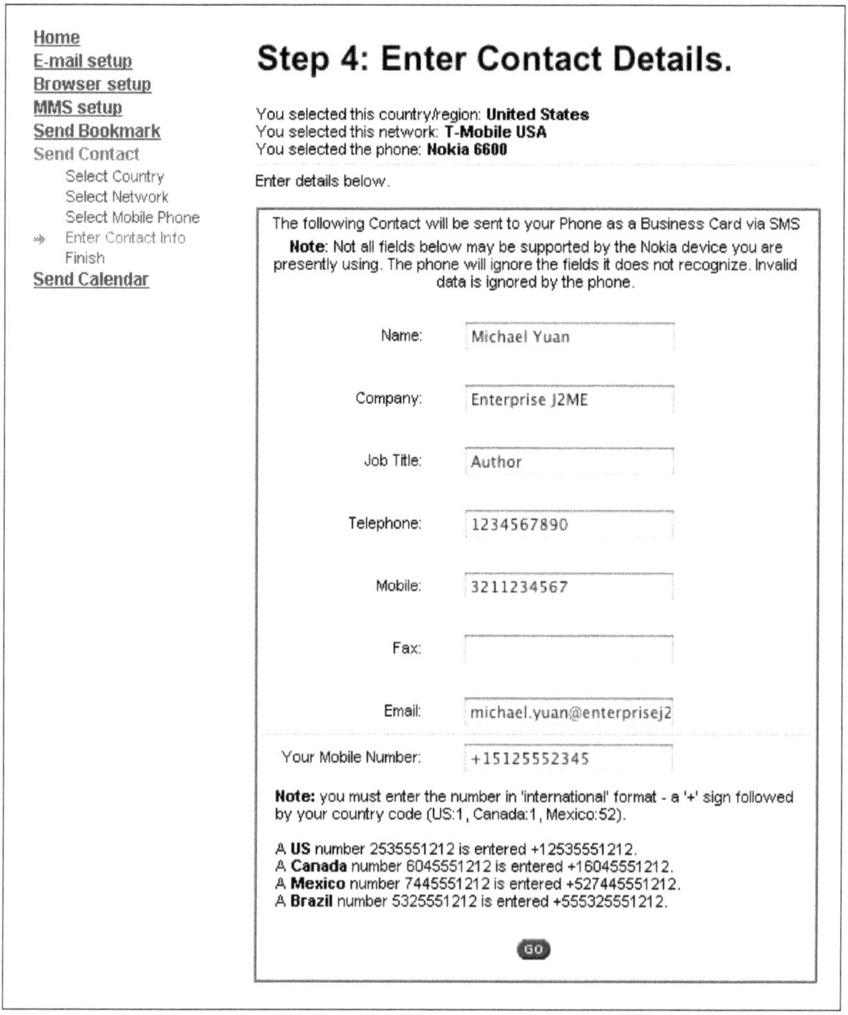

Home
E-mail setup
Browser setup
MMS setup
Send Bookmark
Send Contact
   Select Country
   Select Network
   Select Mobile Phone
⇒ Enter Contact Info
   Finish
Send Calendar

Step 4: Enter Contact Details.

You selected this country/region: **United States**
You selected this network: **T-Mobile USA**
You selected the phone: **Nokia 6600**

Enter details below.

The following Contact will be sent to your Phone as a Business Card via SMS
Note: Not all fields below may be supported by the Nokia device you are
presently using. The phone will ignore the fields it does not recognize. Invalid
data is ignored by the phone.

Name: Michael Yuan

Company: Enterprise J2ME

Job Title: Author

Telephone: 1234567890

Mobile: 3211234567

Fax:

Email: michael.yuan@enterprisej2

Your Mobile Number: +15125552345

Note: you must enter the number in 'international' format - a '+' sign followed
by your country code (US:1, Canada:1, Mexico:52).

A **US** number 2535551212 is entered +12535551212.
A **Canada** number 6045551212 is entered +16045551212.
A **Mexico** number 7445551212 is entered +527445551212.
A **Brazil** number 5325551212 is entered +555325551212.

GO

Figure 6-14. Sending a business card to a phone via a web portal

Options → Send bus card on a Series 40 device. You can choose from several data connection options to send the business card (see Figure 6-15).

The data transport options for sending business cards from the phone are as follows:

- The "Via text message" option composes an SMS message containing the text contents of the business card. The SMS message can only be sent to another mobile phone with a valid phone number. The recipient can view the business card in the message Inbox and save it. But the thumbnail images are lost.

Figure 6-15. Sending a business card from a Nokia phone

- The "Via multimedia" option creates an MMS message with the business card encoded in a file attachment. The thumbnail image, if available, is included in the file. The message can be sent to a phone number or any email address.

- The "Via e-mail" option is the same as the "Via multimedia" option, except that it can be sent only to an email account. But it does not incur the per-message service fee many operators charge for MMS messages.

- The "Via Bluetooth" and "Via infrared" options beam the business cards, including the thumbnail images, to a nearby device. If the recipient is a Nokia device, you can follow the instructions for receiving a business card over Bluetooth from a Mac in "Transfer Contacts from a Computer" [Hack #34]. If the recipient is a PC or a Mac, the business card appears as an incoming file in *.vcf* format with embedded thumbnail data.

Bluejacking [Hack #22] is a prank that uses Bluetooth to send images or messages disguised as business cards to strangers in a crowd. The Nokia Sensor application [Hack #11] helps you identify potentially interesting people in a crowd.

Use the SIM Card

If you want to transfer several phone numbers from one Nokia phone to another, you can just save each into the SIM card, swap the card to the other device, and then copy the SIM contacts to the phone's Contacts list. Figure 6-16 shows how to save a phone number from the Contacts application to the SIM card. Note that you must open the contact and select a phone number first—this will not work if you've selected only a contact from the Contacts list.

You can also speed-dial **[Hack #26]** contacts that are stored on your SIM card.

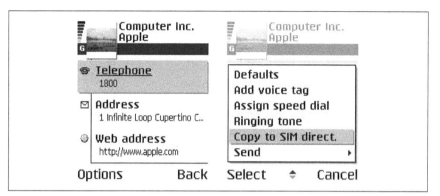

Figure 6-16. Saving a phone number to the SIM card

 A SIM card stores only 250 contacts. Each SIM contact can have only one name and one phone number (this is why you must select one phone number before copying it to the SIM directory). No street addresses, email addresses, notes, or thumbnail images are stored on the SIM card.

Figure 6-17 shows how to browse the phone numbers on the SIM card and save a SIM card entry to the Contacts list. On older Series 60 phones, such as the Nokia 3650, you must open the SIM directory application (select SIM → SIM direct. from the Main menu) to work with the contacts on your SIM card.

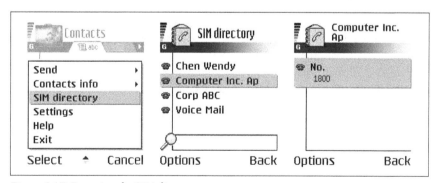

Figure 6-17. Browsing the SIM directory

The SIM card is especially useful when you upgrade to a new phone. You store your most important set of phone numbers to the SIM card. When you upgrade your phone, simply insert the SIM card into the new phone. The new phone now has the same phone number as the old phone, and it has access to the Contacts list from the old phone.

Synchronize Phone Data with a PC

Synchronize the Contacts list, calendar, to-do list, and email messages on a Nokia device with PIM applications on PCs.

When PDAs first came out in the 1990s, their killer feature was the ability to synchronize PIM application data between the mobile device and the PC. PIM data typically includes contact business cards, calendar items, to-do lists, and email messages. A typical synchronization workflow is as follows: the user adds and updates PIM data directly from the PC using the full-screen keyboard. Then, she simply synchronizes all the changes to the mobile device and is ready to go. While on the move, she makes changes to the PIM data on the device (e.g., to add new business cards, receive email messages, or update the calendar). At the end of the day, she simply synchronizes all the changes made on the mobile device back to the computer.

Synchronization is a two-way updating process. It merges the changes made on both ends of the synchronization (e.g., the computer and the mobile device) since they were last synchronized. If an item is updated on both the computer and the mobile device, a conflict can emerge. In this case, the synchronization program prompts you to resolve the conflict. Or, the conflict could be resolved according to a predefined policy. For example, you could specify that the synchronization program always use the version on the mobile phone when there is a conflict.

Synchronization is typically initiated by the desktop computer and processed by the synchronization software on the computer. The computer acts as the main repository of all your PIM data. It backs up the data in the mobile device and supports advanced PIM management features not available on small devices. In addition, mobile devices can synchronize with each other by using a computer as a common repository.

In "Transfer Contacts from a Computer" [Hack #34], I discussed methods to exchange individual contact items between devices and desktop computers. Synchronization automates the process by tracking the changes on multiple devices at the same time.

The Nokia PC Suite contains a Synchronise program, which helps to synchronize contacts, calendar events, to-do notes, and email messages from your phone to the Microsoft Outlook or IBM Lotus applications on a Windows PC. Figure 6-18 shows the synchronization in progress.

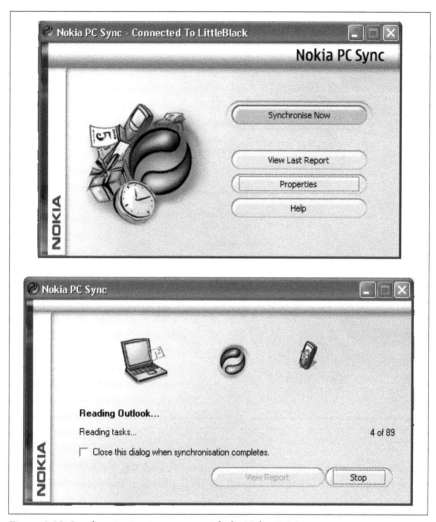

Figure 6-18. Synchronization in progress with the Nokia PC Suite

Figure 6-19 shows how to configure the synchronization profile and policies to set up multiple phones with multiple PIM databases.

 ## Synchronize Phone Data with a Mac

Synchronize your Nokia Series 60 smartphone with a Mac OS X computer.

The Nokia PC Suite does not run on Mac computers. Fortunately, iSync, the default data synchronization program on Mac OS X, supports synchronization with some Nokia Series 60 devices. Please note that you cannot synchronize Series 40 devices with a Mac computer via iSync. Please refer to

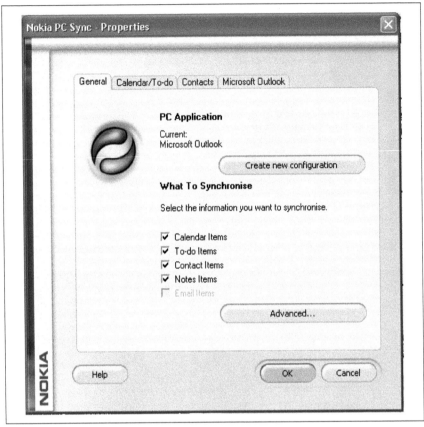

Figure 6-19. Configuring the synchronization profile

"Synchronize Phone Data with Any Computer" **[Hack #38]** for generic options to synchronize any computer with Nokia phones, including Series 40 phones.

> You can find a complete list of devices supported by iSync at *http://www.apple.com/isync/devices.html*. Even if your device is not listed, it is still possible to hack iSync configuration files to get around artificial limitations imposed by Apple. For instance, to set up an unsupported Nokia 9300 with iSync, visit this site: *http://www.macosxhints.com/article. php?story=20050422125043439*.

To use the iSync program, first you should make sure your phone is connected to the Mac computer. I recommend a Bluetooth connection for Nokia phones. The iSync program's main window displays all the mobile devices that are currently being synchronized with the Mac. For example, the list could include all paired Nokia Series 60 phones and all connected iPod devices. Click any of the device icons to configure their synchronization settings (see Figure 6-20). If your device is paired but it does not appear in the list, you can use the Devices → Add Device menu to search for it and then add it.

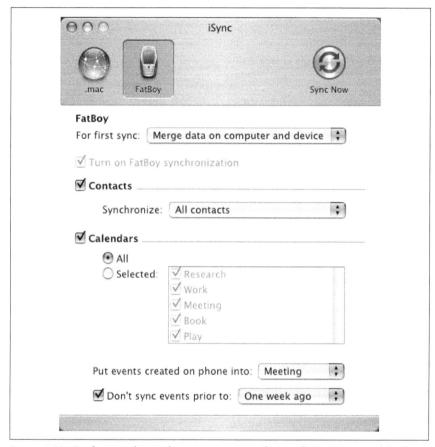

Figure 6-20. Configuring the synchronization options for a Nokia Series 60 mobile phone

You can choose to synchronize any of the listed devices by clicking it and then clicking the Sync Now button. If you click the Sync Now button without choosing a device, it goes through all the listed devices one by one and tries to synchronize them. If a device is not available, iSync silently moves on to the next device, but gives you an error message at the end. Figure 6-21 shows the synchronization in progress.

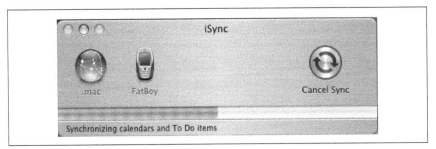

Figure 6-21. Synchronization in progress

> The iSync program uses TCP/IP port 3004 to communicate with the phone. Make sure you have that port open if you have an active firewall on the Mac.

For Nokia Series 60 devices, iSync synchronizes all the contacts (including thumbnail pictures) to the Mac Address Book program, and synchronizes all calendar items and to-do lists to the iCal program. Email messages are not synchronized.

> Using a Nokia Series 60 phone as a bridge device, you can synchronize a Windows PC and a Mac computer. This is a three-step process. First synchronize the PC with the phone [Hack #36], then synchronize the phone with the Mac, and then synchronize the phone with the PC again. The first two steps guarantee that the changes on the PC are passed to the Mac, and the last step ensures that changes in the Mac are synchronized back to the PC.

Synchronize Phone Data with Any Computer
#38
SyncML allows you to synchronize a Nokia phone with almost any PIM database on computers and servers.

The Nokia PC Suite and Apple iSync both initiate the synchronization session from the computer over a local network connection (e.g., a Bluetooth connection). But many times those are not sufficient. Here are some examples:

- The client-based solutions do not allow mobile phones to directly synchronize with popular workgroup servers, such as the Microsoft Exchange server.

- Few Nokia-compatible synchronization programs are available for different flavors of Linux and Unix desktop computers.

- Apple iSync does not synchronize any Series 40 device with a Mac computer. It also does not support all popular Series 60 devices.
- The Nokia PC Suite does not always work properly for all PCs and all phones.

 Please refer to "Synchronize Phone Data with a PC" **[Hack #36]** for the basics of mobile data synchronization.

SyncML, a generic XML-based language for expressing data exchange in a synchronization session, provides the answer. Many Nokia devices are capable of initiating synchronization sessions to SyncML servers via general Internet connections (e.g., TCP/IP over GPRS). A SyncML server communicates with several synchronization endpoints over TCP/IP connections using SyncML. Most commercial PIM databases support SyncML. Check your Nokia phone's manual to see if it supports SyncML.

 If your Series 60 phone does not have built-in support for SyncML (e.g., Nokia 3650), it is still possible to add SyncML support by installing new native software on the phone. Nokia provides Symbian-based SyncML software for the Nokia 3650 free of charge (see *http://www.nokia.com/nokia/ 0,8764,5371,00.html*).

The SyncML servers enable the device to synchronize with a large variety of backend data sources that do not have native Nokia support. Here are three examples:

Run a personal SyncML bridge

The open source MultiSync program synchronizes a Nokia device with the Ximian Evolution (or Evolution 2) PIM software on any GNOME-based Linux system. You can download MultiSync from *http:// multisync.sourceforge.net/*. Then you need to run MultiSync on your personal Linux computer, together with GNOME Evolution. The mobile phone connects to MultiSync via GPRS Internet and synchronizes with Evolution through MultiSync. A tutorial on how to configure Multi-Sync with a Nokia 6600 is available in the wiki knowledge base at *http:// multisync.sourceforge.net/wiki/index.php?Nokia6600Instructions*.

Run a dedicated SyncML gateway server

The open source Sync4j Project (*http://www.sync4j.org/*) and Synchronica (*http://www.synchronica.com/*) provide software for dedicated SyncML gateway servers. They synchronize Nokia devices with workgroup-based PIM servers, such as the Microsoft Exchange server. You can run the gateway SyncML server to provide synchronization services to many mobile users. Data in the Microsoft Exchange servers can then be synchronized to almost any Windows or Mac computer using native synchronization software.

Hosted SyncML services

FusionOne's MightyPhone service (*http://www.mightyphone.com/*) is a hosted SyncML server that synchronizes mobile phones with Outlook or Lotus software on a Windows PC. To use the service, you need to open an account on the FusionOne web site for $3 per month (a free trial is available). Both the mobile device and the PC synchronize with the hosted account. Hence, the account always stores the updated PIM data and acts as the middleman to propagate changes from the phone to the PC or vice versa. For instance, if you make a change on the phone and synchronize it to your MightyPhone account, the change is propagated to the PC the next time the PC software synchronizes with MightyPhone. When you make changes on the PC and synchronize them to your MightyPhone account, the service sends an SMS message to the phone to alert you to initiate synchronization to keep the phone up-to-date.

On a Nokia Series 60 device, the synchronization process is handled by the Sync application (accessible from the Connect → Sync menu). The Sync application holds one or multiple synchronization profiles. Each profile contains the settings for the SyncML server, the data set to be synchronized, the target databases, and the synchronization policy. You can create a new profile for your SyncML server or edit existing profiles via the Options menu. Figure 6-22 shows the SyncML server connection settings including the IP address, TCP/IP port number, and user authentication credentials. It also shows that contacts and calendar data are to be synchronized. The Remote Database values are specific to the SyncML server. In this case, the "address-book" and "calendar" values refer to the GNOME Evolution address book and calendar programs managed by a MultiSync instance.

To start the synchronization process for the selected profile, you can use the Options → Synchronise menu item.

Read PC and Mac Documents

#39   Sure, the screen's small and you might have to scroll and squint a bit, but if you've got to read documents while you're on the go, something's better than nothing!

Working with office documents while on the run might seem tedious at first. Most mobile phones have keypads that don't exactly lend themselves well to

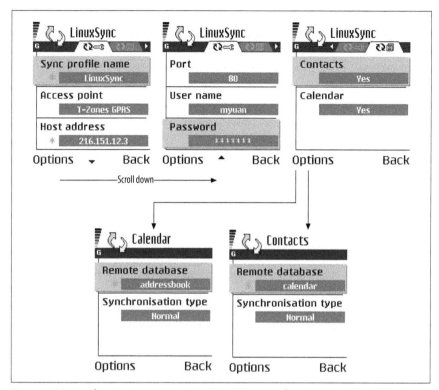

Figure 6-22. Configuring connections to a SyncML server from a Series 60 mobile phone

text input, but I've often found myself wanting to read them, especially when I receive attachments via email. With the help of some affordable software, you can view, and in some cases even edit, documents that start out life on a PC or Mac.

I have been using Quickword Viewer ($14.95 from *http://www.quickoffice. com/*) to view documents while they are sitting in the Inbox on my Nokia 6600, waiting to end up on my computer when I get back home or into the office. Quickword also can read Palm DOC eBooks. Quickword is fast, and can run at full screen to show the most that any document viewer really can show on a Series 60 device. Quickpoint Viewer ($14.95) is also available, and will let you view PowerPoint documents on your phone.

You can take things a step further with Quickoffice Premier ($49.95), which can read and write Excel, Word, and PowerPoint files. However, it requires a newer Series 60 device. If you are using an older device, such as the Nokia 3650, you'll be limited to Quickword and Quickpoint. Figure 6-23 shows a Word document in Quickoffice Premier.

Figure 6-23. Opening a Word document in Quickword

RepliGo (*http://www.cerience.com/*) is another option. It syncs your documents back and forth, similar to the Documents to Go application available to Palm OS users. RepliGo requires a PC to perform conversions, leaving Mac users totally in the cold. (Though like many Mac users, I'm used to getting kicked in the stomach by software developers now and then, so this doesn't really bother me much.) The RepliGo viewers are available for free, which is good news for folks who spend money on the converters, since it makes it possible for anyone with a supported device to view the converted office documents.

Like RepliGo, the Mobipocket Office Companion (*http://www.mobipocket.com/*) converts your office documents to a format that you can view on your phone. The reader is free, but you'll need to buy the converter ($19.95; $29.95 adds Access, FrontPage, and Visio support).

PDF Documents

The PDF format is a great way to share documents on a variety of devices. Adobe Reader (*http://www.adobe.com/*) exists for desktop computers and Palm OS devices. You can view PDF files on a Series 60 device by using Pdf+ from mBrain (*http://www.mbrainsoftware.com/Pdf/Pdf.htm*).

Installation is a snap, of course, and you can open PDF documents right from the Inbox of your handset's Messaging program. You can also store a PDF document on the MMC or Memory Stick and open it up for viewing.

 Nokia 6680 users can get Acrobat Reader directly from Adobe. Visit Adobe's site at *http://www.adobe.com/products/ acrobat/readerforsymbian.html* to learn more.

Pdf+ costs $25 from Handango (*http://www.handango.com/*) and is a great way to view PDF documents on your mobile phone. I have experienced a couple of instances in which Pdf+ certainly earned its keep.

MapQuest (*http://www.mapquest.com/*) has "printable" directions that print easily to PDF files from Mac OS X (or via other means on Windows PCs), and scale well in most circumstances to the display on a Nokia 6600. I get the directions, save them as a PDF, and send them to my mobile phone using Bluetooth, and I'm out the door with directions to my destination. Figure 6-24 shows how the directions from Providence, R.I., to Hell, Mich., look in Pdf+, and Figure 6-25 shows how they look with text wrapping enabled (12 hours... that's a pretty short trip).

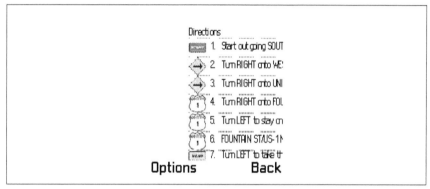

Figure 6-24. Viewing directions from Providence, Rhode Island to Hell, Michigan

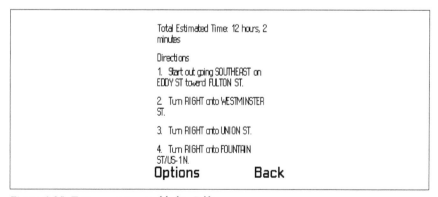

Figure 6-25. Text wrapping enabled in Pdf+

Documentation for your mobile phone might be available in PDF format, and what's better to read than the manual while you're waiting for your stop on the subway or bus? You might learn something cool about your phone while you're at it. You can also save various eBooks to the PDF format however you desire, and stash a copy on your MMC card for reading later.

PDF is not the only format for electronic books. Some older books are available in plain-text format. Or, you can save your Word documents to plain text as well. To view long text files on a mobile phone, try the open source Mobile Bookshelf at *http://bookshelf.sourceforge.net*. It parses the long text into multiple pages for easy viewing, and it supports the full-screen reading mode. Mobile Bookshelf is a Java program that runs on both Series 40 and Series 60 phones.

Hacking the Hack

Instead of running helper applications on your Nokia, wouldn't it be great if your email and web applications had native support for the documents you want to view? Reqwireless's EmailViewer [Hack #60] and WebViewer [Hack #50] do just that. Actually, it's not that these applications know how to display PDFs and Word documents. Instead, the Reqwireless proxy service ($5.99 quarterly, or a $19.99 one-time fee) translates the documents on the fly, displaying documents right in your web browser. Figure 6-26 shows a portion of one of the Word documents from the first draft of the book you're reading right now.

Figure 6-26. Reading a Word document in WebViewer

WebViewer and EmailViewer can look inside zip files, and will display Microsoft Word and Excel, Corel WordPerfect, and PDF documents. Not only that, both programs will run on practically any Java-powered phone, so it's a perfect choice for both Series 40 and Series 60 users.

—Emory Lundberg

Enhance the PC Experience
Hacks 40–42

It is easy to understand how a companion computer can add value to a mobile device. But how about the other way around? Your mobile phone can make your computer more useful by providing ubiquitous access to the Internet via your mobile data service. You also can use your phone as an alternative interface to remotely control the computer.

 HACK #40 ## Connect Your Computer to the Mobile Network

Get almost-free Internet access to your laptop at any time, from anywhere, using your mobile phone as the data modem. No WiFi coverage is needed!

As a business traveler, I've always longed for Internet access on the road—not only for my mobile phone, but also for my laptop. After all, it's much more convenient to compose email and browse the Web on a laptop than on a phone. A killer mobile phone application for me is to use the phone as a data modem to hook my laptop to the Internet anytime, anywhere. It complements public WiFi networks, and helps me to get around the limitations of WiFi hotspots!

While WiFi networks have made great progress in providing public wireless Internet access, the mobile phone data network still offers several notable advantages when it comes to business users. First, WiFi hotspot coverage is still limited. For instance, typically there is no WiFi coverage in parks, in many government facilities, or at highway rest areas. The mobile phone data network, on the other hand, is ubiquitous. It is available in most cities and along major highways. This always-on feature is a major selling point for business travelers.

Even in places with WiFi coverage, the network is often commercial and requires a per-use fee. For instance, different commercial WiFi networks might be installed in the coffee shop around the corner, in the bookstore, in the hotel, and in the airport. So, if you travel to several of these places in one day, you might need to pay tens of dollars in WiFi access fees. The mobile phone data network, on the other hand, is operated by national operators. You know exactly what you need to pay for data access each month.

Today's wide area mobile data networks **[Hack #4]** are built on several different technologies. Which network is available to you depends on your device, location, and service plan. In general, the faster data networks (e.g., UMTS, 1xEV-DO, and EDGE) are more expensive to use and more limited in coverage when compared with slower and more ubiquitous networks, such as General Packet Radio Service (GPRS).

Most mobile network operators offer special data service plans for laptop computers these days. Those plans give you prorated bandwidth of wireless data, or even unlimited access for a flat monthly fee. You can simply purchase a GPRS or EDGE wireless card and plug it into your laptop. The wireless card requires a valid SIM card from the operator to authenticate the laptop to your service account on the network. Those special data plans do not include any voice minutes.

> As an alternative, you can do a SIM-swap with your mobile phone, if you have a data plan that is designed to work with PC cards. For example, T-Mobile's inexpensive, unlimited data plan ($20 per month with most voice plans, $30 otherwise) will work great with the very low-cost Merlin G100 data card, which you should be able to find on eBay for less than $50. Although the companion software is Windows-only, most Merlin G100s will work out-of-the-box with Mac OS X and Linux, appearing as serial ports that you can configure to dial the same number (*99#) you'd use with your phone, as shown later in this hack.
>
> When you want to dial out using the data card, swap the SIM card out of your phone and into the card, and swap it back when you're done. As an added bonus, no one will be able to bother you by phone while you're online!

However, if you are a casual and budget-minded wireless data user like I am, the dedicated wireless card and a data-only service subscription, in addition to your voice service, are probably a little too expensive. Instead, you can simply add GPRS or EDGE Internet service to your existing mobile phone service plan and then share the connection from the phone to the laptop computer. This is also known as *tethered* Internet access. Once connected, you should be able to use the laptop to browse the Web, check email, and telnet/ssh/ftp to other sites.

Which Access Point to Use?

As I discussed in "Connect Your Phone to the Internet" [Hack #10], a mobile phone can connect to the Internet via several different connection (a.k.a. access point) configurations. You need to specify which connection to use when you're using the phone as a modem. In most cases, you can select the access point using the phone-connection software on the computer (see the instructions in the next several sections). Or you can set the active access point on the phone via the Tools → Settings → Connection → GPRS → Access Point menu on a Series 60 device, or the Settings → Connectivity → GPRS → GPRS Modem Settings menu on a Series 40 device.

To share the network from the phone to the computer, you need to subscribe to an Internet data plan that does not require WAP gateways (most handset data plans will support WAP, HTTP, and email, but little more than that). Remember to select the Internet access point for that plan for the phone modem.

Use the Nokia PC Suite

On a Windows PC, you can use the Nokia PC Suite to connect to the phone as a GPRS modem. The Nokia PC Suite probably already installed the modem driver for the phone for you. If it didn't (or if you add more phones to the PC Suite), you can download the Windows modem driver for most Nokia phones via the product software download page on the Nokia web site. The driver file has a *.inf* suffix. You should right-click it and choose Install from the pop-up menu to install it.

Once the modem driver is installed, you need to add the mobile phone as a modem to the Windows system. Open the Phone and Modem Options icon in the Control Panel and click the Modem tab. Click the Add... button to add a new modem. You should skip the modem detection step and select a Nokia phone modem from the list (see Figure 7-1).

Figure 7-2 shows that the phone modem is installed on an available COM port in the Phone and Modem Options configuration.

In the Nokia PC Suite, open the Edit Modem Options program and select a GPRS access point for this modem (see Figure 7-3).

Now, you can add a new Internet connection for this new modem in the Network Connections section of the Control Panel (see Figure 7-4). The Windows New Internet Connection Wizard asks you the phone number the modem should connect to—try the phone number *99# for GPRS and EDGE connections (use #777 for Code Division Multiple Access, or CDMA, data service). That's it! Now you can use the new network connection to connect to the Internet—it automatically dials and connects the phone behind the scenes.

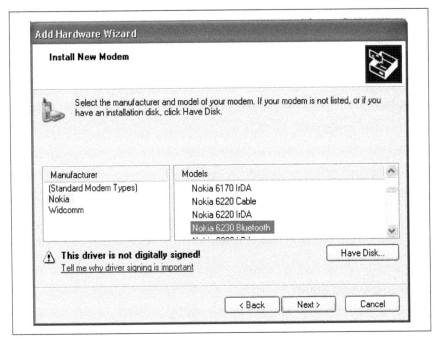

Figure 7-1. Adding a mobile phone modem to the PC

The process I just covered is pretty complex and is difficult to troubleshoot if anything goes wrong. To make life easier for users, Nokia provides a free download called One Touch Access (*http://www.nokia.com/nokia/ 0,8764,72028,00.html*). It allows you to share an Internet connection from your phone, with just one touch. The One Touch Access program requires Nokia PC Suite v6.4 and above.

If you have difficulty working with the Nokia PC Suite to share an Internet connection, try QuickLink Mobile for Windows, from Smith Micro (*http:// www.smithmicro.com/*). This is a nice utility for automating and managing Internet connections on a Windows computer, and it helps you to set up a mobile phone data modem. A Mac OS X version of QuickLink Mobile is also available (I discuss it later in this hack).

Use Bluetooth on the Mac

On a Mac OS X computer, first you need to configure the Bluetooth connection to support direct GPRS dialing. Open the Bluetooth Setup Assistant program from the Bluetooth Control Panel in System Preferences, or from the Bluetooth icon in the system menu. Figure 7-5 shows that you should select the GPRS direct connection option in the pairing wizard and set the connection phone number to *99#. The connection script is Nokia Infrared

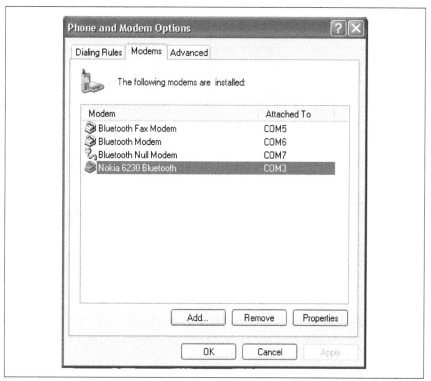

Figure 7-2. The Bluetooth modem installed

for most Nokia mobile phones. But you can also experiment with others if you get a connection error.

Ross Barkman maintains a web page with a comprehensive list of tips and dialing scripts for connecting a Mac to the Internet via a mobile phone modem. Ross's scripts allow you to specify the access point on the Mac. You should definitely visit this web page if you encounter any connection problems. The web site URL is *http://www.taniwha.org.uk/*.

Once the Mac and phone are set up correctly, you can use the Internet Connect program to connect the Mac to the Internet via the phone modem. After you click the Connect button, the phone displays an alert message. You must explicitly confirm that message on the phone before any actual GPRS connection is made. Once the Mac is connected, the Internet Connect window displays the incoming and outgoing data traffic (see Figure 7-6).

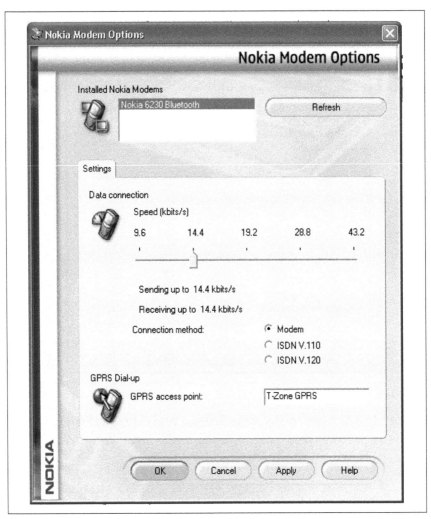

Figure 7-3. Configuring modem access in the Nokia PC Suite

Again, the QuickLink Mobile for Mac OS X software from Smith Micro (*http://www.smithmicro.com/*) provides a simple graphical user interface (GUI) to manage all Internet connections, including wireless phone data modems, on a Mac computer. It supports Internet sharing across both Bluetooth and cable-based connections to the phone. Smith Micro even sells Mac-compatible Nokia data cables, since most standard cables are PC only.

Connect a Linux Computer

On a Linux computer, the mobile connection via GPRS requires several setup steps. First, if your phone is connected to the Linux computer via

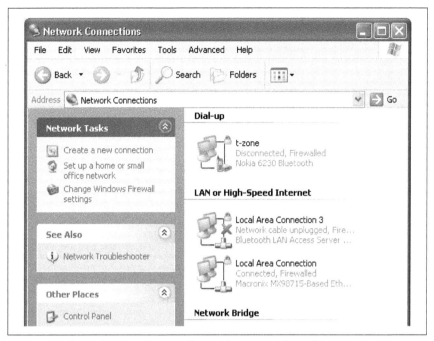

Figure 7-4. The modem connection in the Windows Control Panel

Bluetooth, you need to bind the RFCOMM service to a specific serial port. The following commands first find the Bluetooth ID of the connected phone, then determine that RFCOMM service is available on channel 3, and then bind the RFCOMM connection to the PC's serial port number, 0:

```
$ sdptool browse
00:60:57:50:AB:9C

...
Service Name: Dial-up Networking
Service Description: Dial-up Networking
Service Class ID List:
  "Dialup Networking" (0x1103)
Protocol Descriptor List:
  "L2CAP" (0x0100)
  "RFCOMM" (0x0003)
    Channel: 3

# rfcomm bind 0 00:60:57:50:AB:9C 3

# rfcomm
rfcomm0: 00:60:57:50:AB:9C channel 3 clean
```

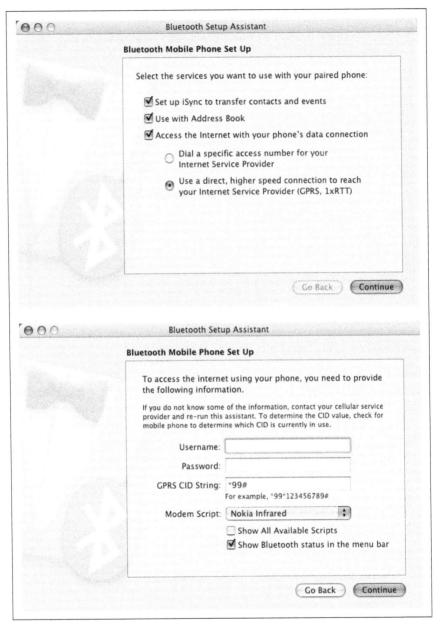

Figure 7-5. Configuring the Mac and the phone to share a GPRS data connection

With the RFCOMM connection bound to the */dev/rfcomm0* device, you can use the following script to actually connect the Linux computer to the Internet using the PPP protocol over the phone's GPRS connection:

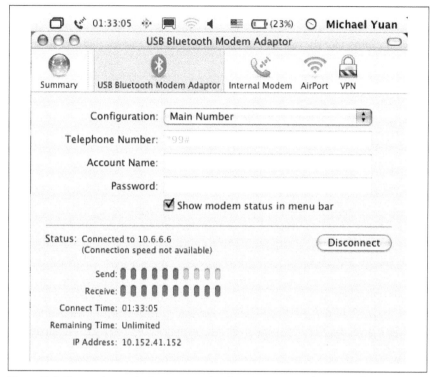

Figure 7-6. Connecting to the Internet via the phone's GPRS modem using the Internet Connect program

```
# File: /etc/ppp/peers/btnokia
#
/dev/rfcomm0  # The Nokia phone
115200        # speed
defaultroute  # use the cellular network for the default route
usepeerdns    # use the DNS servers from the remote network
nodetach      # keep pppd in the foreground
nocrtscts     # no hardware flow control
lock          # lock the serial port
noauth        # don't expect the modem to authenticate itself
local         # don't use Carrier Detect or Data Terminal Ready

connect    "/usr/sbin/chat -v -f /etc/chatscripts/connect"
disconnect "/usr/sbin/chat -v -f /etc/chatscripts/disconnect"
```

Please refer to "Configure Bluetooth for Linux" **[Hack #14]** for details on how to set up a Bluetooth connection between a Nokia phone and a Linux computer.

Of course, you do not have to use Bluetooth to connect the phone to the
Linux computer. If you use the USB data cable, the connection script could
look something like the following:

```
# File: /etc/ppp/peers/usbnokia
#
/dev/ttyUSB0   # USB-serial port to the Nokia phone
230400         # speed
defaultroute   # use the cellular network for the default route
usepeerdns     # use the DNS servers from the remote network
nodetach       # keep pppd in the foreground
crtscts        # hardware flow control
lock           # lock the serial port
noauth         # don't expect the modem to authenticate itself

connect    "/usr/sbin/chat -v -f /etc/chatscripts/connect"
disconnect "/usr/sbin/chat -v -f /etc/chatscripts/disconnect"
```

The *connect* and *disconnect* scripts at the bottom of the preceding two scripts
determine how the phone communicates with the network to establish the
GPRS connection. The access point name (APN) is specific to the wireless
operator. Here is an example *connect* script for the AT&T Wireless Service
(replace *proxy* with the APN for your provider; for a comprehensive list of
providers and their APNs, see *http://www.opera.com/products/mobile/docs/
connect/*):

```
TIMEOUT 10
ABORT   'BUSY'
ABORT   'NO ANSWER'
ABORT   'NO CARRIER'
SAY     'Starting GPRS connect script\n'

# Get the modem's attention and reset it.
''      'ATZ'

# E0=No echo, V1=English result codes
OK      'ATE0V1'

# Set Access Point Name (APN)
SAY     'Setting APN\n'
OK      'AT+CGDCONT=1,"IP","proxy"'

# Dial the number
SAY     'Dialing...\n'
OK      'ATD*99***1#'
CONNECT ''
```

The *disconnect* script for the AT&T Wireless Service is as follows:

```
""      "\K"
""      "+++ATH0"
SAY     "GPRS disconnected."
```

To make the call, you can use the pppd call *peer* command, where *peer* is the name of the peers file (*btnokia* or *usbnokia* in the earlier examples).

For more *connect* and *disconnect* scripts for other wireless operators, please refer to the book *Linux Unwired* (O'Reilly, 2004).

Getting Around Operator Limitations

Wireless network operators often impose constraints on how you can use their data networks. For instance, most operators allow only network traffic via TCP/IP ports 80 (web pages) and 25 (email) to get through to the phone. All other ports are closed by filters at the operator's Internet access point server. However, on premium services, such as Cingular's $80/month unlimited data plan, or T-Mobile's $20–$30/month offerings, you are unlikely to find that the traffic is restricted. However, you might need to use a different APN (for example, T-Mobile uses the *internet2.voicestream.com* and *internet3.voicestream.com* APNs for its unlimited data users). If you're going to rely on your cell phone's data plan as your lifeline, it's well worth it to sign up for an unlimited plan that supports access via your notebook computer or PDA. Otherwise, you're taking chances with a lower tier of service that promises much less than what you need.

To access data services via other ports, you can set up a gateway server for the phone to channel generic Internet traffic through the permitted ports on the phone. One such solution is to set up a Virtual Private Network (VPN). A VPN also encrypts communication data over open channels and provides secure data access to corporate networks. You need to configure the VPN to listen on one of the supported ports (80 or 25) and configure your client software accordingly. To learn more about how to set up a VPN for Windows PC, Mac, and Linux computers, please refer to *Virtual Private Networks*, Second Edition (O'Reilly, 1998).

A Remote Control for Your PC

#41

Control media playback and PC applications from your phone.

One of the coolest mobile phone hacks is to use the phone as a remote control for your PC. Why would anybody want to do that? Isn't the keyboard and mouse enough to "control" the PC? Well, as it turns out, there are at least a couple of reasons you'd want to control your PC via a wireless device:

- You can use the PC to run PowerPoint presentations, software demonstrations, and/or photo slideshows for an audience. To capture the audience's attention and convey your messages effectively, it is important that you control the presentation PC remotely while you are walking across the room. Dedicated presentation remote control device are available for purchase; however, the mobile phone eliminates the need for those separate gadgets.

- With a remote control, you can turn the PC into a fully functional media player in your living room. The PC plays DVD, CD, MP3, and many other media formats. More importantly, you can update the PC software to support future media formats. The major obstacle that currently hinders the adoption of media PCs in living rooms is the lack of a good remote control—a keyboard and mouse attached to the media console are simply not acceptable.

Wireless local connectivity technology such as infrared or Bluetooth is included in most Nokia mobile phones to allow the transfer of data between devices over short distances. Phone applications can use this same technology to remotely control computers. Unlike traditional remote controls for TVs and presentations, smartphones are generally much higher-quality devices with high-resolution, full-color displays. This opens up a whole host of possibilities when it comes to interactivity and the overall user experience.

Using Bluetooth allows you to communicate with devices without requiring direct line-of-sight. So, unlike with an infrared remote control, you can be in another room and still control your device. Bluetooth connections also allow for bidirectional communication of data at higher speeds than infrared, and they support the option of being encrypted to prevent anyone nearby from "listening in" on what you are doing.

Control Your PC for Media Playback

The best way to control media playback is to have the media files stored on your computer and to use your phone as the controller and display device. The phone reads your commands, and transmits them to a media playback program on the PC to perform such actions as playing, pausing, and skipping songs. Several popular media players are available for the PC. The open source Winamp (*http://www.winamp.com/*) program is the best, since it is freely available to anyone and you can extend it with custom plug-ins. You can write (or obtain) Winamp plug-ins to interact with the remote control software on the phone.

Bemused and ControlFreak. Not long after the first Series 60 smartphone was released (Nokia 7650 in June 2002), a couple of new products were developed that took advantage of the Bluetooth capabilities of the new phone and the music storage and playback of a PC.

Bemused was written by Ashley Montanaro and was released as open source in 2002. You can download the Bemused software, including the Symbian client for the phone and the Winamp plug-in on the PC, from *http://bemused.sourceforge.net/*. Bemused provides the usual media player options in the form of on-screen buttons in a skinnable interface. It lets you control the current playlist as well as send the entire directory and file structure of your music to the phone. This can take quite a long time if your music collection is large, but once the data is received, it allows basic browsing and loading of your music. Bemused supports both GPRS and Bluetooth connections. With GPRS, you can control your music remotely via the Internet. However, you should be aware that most mobile operators charge for GPRS data, whereas using the local Bluetooth connection is free and more reliable.

Around the same time Bemused became available, a user interface (UI) design engineer, Michael Ahokas, and a software design engineer, Trent Fitzgibbon, were developing a commercial application called ControlFreak for a similar purpose. But compared to Bemused, ControlFreak has a much more polished UI and richer functionality. For instance, most ControlFreak operations are performed with a single click of the navigation pad (i.e., the joystick). You can also set ControlFreak to display track information in any application, and you can use shortcuts to adjust the volume or skip tracks from the phone application idle screen. We believe that usability is the key to any phone-based application. So, we will focus on ControlFreak in this hack. Figure 7-7 compares the player UI screens of Bemused and Control-Freak.

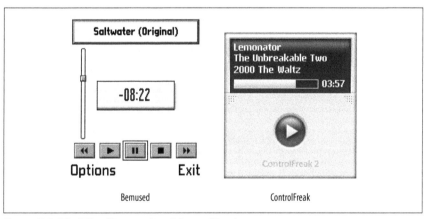

Figure 7-7. The UIs of Bemused and ControlFreak

 Both Bemused and ControlFreak are Symbian applications, which run only on Series 60 devices. A Java port of Bemused called *Bemused.java* is available from *http://elektron.its. tudelft.nl/~jkohne76/*. It runs on any Nokia mobile phone with Java Bluetooth API support (e.g., the Nokia 6230 phone in Series 40). But *Bemused.java* is a much less-polished product than Bemused and ControlFreak.

Use ControlFreak for media playback. You can download ControlFreak from *http://mtvoid.com/*. It contains a Symbian application for the phone and a Winamp plug-in for the PC (see Figure 7-8). The Winamp plug-in takes commands from the phone via a preconfigured Bluetooth COM port (see Figure 7-9) and controls the PC.

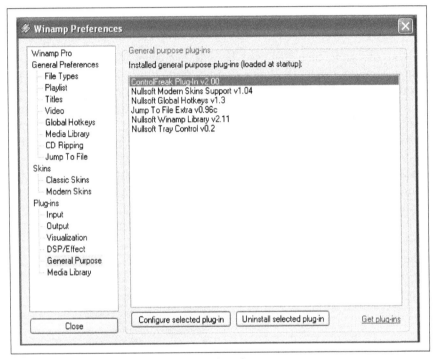

Figure 7-8. The Winamp plug-in for ControlFreak

After installing the program and configuring your Bluetooth connection, you initiate a connection from your phone to your computer using the Connect menu option, or simply by pressing the navigation pad. After searching for nearby Bluetooth devices, you will be presented with a list. The next time you select Connect you don't need to search for devices, since ControlFreak remembers which device you last used and asks if you want to reconnect. In fact, if you enable Reconnect on startup in the settings, ControlFreak will automatically connect the next time you start the application. Once the connection is made, the virtual device cover will slide down to reveal details about the current track.

In the main player mode, the top part of the screen shows details about the current track (including artist, title, album, and year if ID3 tags are present). Below that is either more track information, a preview of upcoming songs,

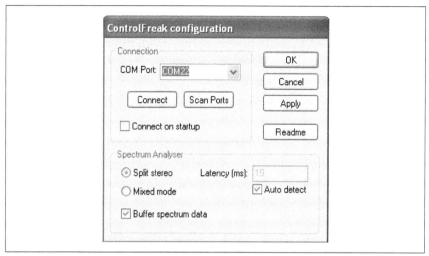

Figure 7-9. Configuring the Winamp plug-in

or real-time spectral analysis of the current audio, depending on the view selected. You can change views using either the Options menu or 1, 2, and 3 shortcut keys. By pressing 0, you can also open or close the cover. Figure 7-10 shows the view for playing back a song.

Figure 7-10. View on the phone remote control when playing back a song

When you are in any of these player modes, you can easily control the playback of your music by using the navigation pad. Pressing the center key on the navigation pad pauses or plays the music. Pressing and holding the navigation pad stops the music. Pressing the navigation pad up or down increases or decreases the volume. Pressing the navigation pad to the right or left skips forward or backward in the playlist. And pressing and holding the navigation pad to the right or left allows you to quickly skip through the current track.

When using your phone, you might not want to have ControlFreak in the foreground all the time. So, to keep track of what is currently playing, the settings have a "Track change" pop-up option. With this enabled, you get a little pop-up box in the top of the screen showing track information changes, no matter what application you are in. You also have the option of enabling "Phone app" shortcuts. With this enabled, you can use basic controls such as play/pause, next/previous track, and volume up/down from the phone application idle screen.

The other modes available enable you to browse your media library, manage the current playlist, and control the desktop. You also can access these from the Options menu or by using their respective shortcuts: 6, 7, and 8. Figure 7-11 shows playlist management on the remote control.

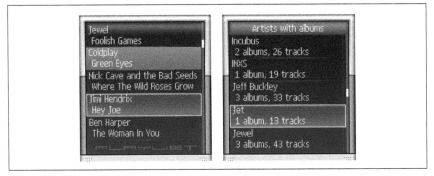

Figure 7-11. Managing the playlist and browsing the media library

The playlist mode displays the contents of the playlist you have loaded in Winamp. You can move the focus up or down by clicking the navigation pad up or down, or speed things up by clicking the navigation pad left or right to move a full screen at a time. The smart thing about the way the playlist view in ControlFreak works is that it loads information from the computer only when it needs it. So, even if you have thousands of tracks in your playlist, you won't be sitting there waiting for the data to be sent to the phone before you can start browsing. If you want to start playing a track, just select it and press the navigation pad. If you want a track to start playing after the current one finishes, just press and hold the navigation pad. The track will flash a couple of times, and then will move to the end of the current track. If you keep repeating this with a number of tracks you can build up a queue of tracks to play. To remove a track from the playlist press the C key or choose Remove from the Options menu.

Searching for a track is a particularly useful feature. You can select Search from the Options menu or just start typing keys, and ControlFreak automatically opens a search query that includes T9 predictive text support. The search results are based on a substring match of the query and the playlist titles (unless the query is only one or two characters). You can also sort the playlist in a variety of ways, and randomize it. This is a better way of playing your music randomly than using your player's shuffle setting, since you are still able to see the track order.

The media library mode allows you to browse all of the audio and video files added to Winamp's media library. You can browse artists, albums, videos, and playlists. When you find something you like, simply select Play or Queue from the Options menu. You can search for individual tracks in your library from the browse mode list, just as you can in playlist mode. You search for a particular item within a list via the jump-to feature. Just type the first letter of what you are looking for, using multiple taps of the number keys, and the focus jumps to the first item that starts with that letter. If you can't think of what to play, you can always just press the 0 key and Control-Freak will jump to a random item.

One of the more exciting features of ControlFreak is the ability to view your desktop directly on your phone (this is called *desktop control mode*). Watching a movie on your computer and having it display the video on your phone is one cool use of this feature. To do this, though, you need to disable hardware video overlay in Winamp's video preferences. By default, it plays video directly to the video card to increase playback performance. Figure 7-12 shows a movie streamed from a PC to a phone screen via ControlFreak.

Figure 7-12. Playing PC video files on the phone

Control Other PC Applications

You also can use ControlFreak to control other PC applications. Winamp plug-ins act as software agents for ControlFreak. Those agents listen for commands from the ControlFreak program running on the handset, and then execute those commands on the PC.

The desktop control mode allows you to do almost anything you normally do with your computer, but from anywhere in its Bluetooth zone. You use the navigation pad to control either the mouse cursor or the keyboard arrow keys; pressing 1 or 2 sends a left- or right-mouse click. It would be very difficult to do anything if the entire desktop was always resized to the phone's screen, so ControlFreak provides zoom functionality. Use 5 to zoom in and 0 to zoom out, just like in the Series 60 image viewer. Useful shortcuts are also provided for sending Enter (7), Esc (9), and Alt-Tab (*) keys. While switching applications, ControlFreak automatically zooms into the switching window to make it easier to see; then you can use the navigation pad to move left or right in the list of applications.

The combination of being able to see what is on the screen and to easily send left/right arrow keys makes desktop control mode a great tool for giving presentations. Showing PowerPoint slides to your customers or photos to your relatives has never been easier. Figure 7-13 shows ControlFreak in the PowerPoint presentation mode.

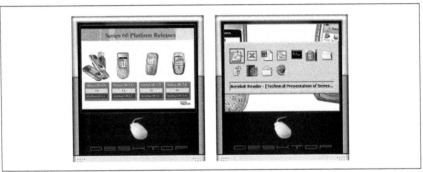

Figure 7-13. Using ControlFreak to run a presentation and switch to other applications

If you're presenting in the dark and you don't want your phone's backlight to keep coming on when clicking buttons, go to the settings and set Backlight mode to Disabled (although this feature is not present in all phones).

—Trent Fitzgibbon and Michael Ahokas

A Remote Control for Your Mac
HACK #42
Use your Nokia phone to control applications on your Mac, including iTunes, iPhoto, DVD Player, KeyNote, PowerPoint, and much more.

ControlFreak is a very useful remote control program for Windows PCs [Hack #41]. But how about Mac users? Mac computers come bundled with beautiful hardware and very powerful multimedia software that make them the perfect entertainment control center at home or on the road. Can we control them remotely via our phones as well? Sure, using a product called Salling Clicker.

Salling Clicker is a Bluetooth-based remote control program for Mac OS X. Written by Jonas Salling, it works on Nokia Series 60 and several other mobile phones. You can download/purchase it from *http://homepage.mac.com/ jonassalling/Shareware/Clicker/*.

Like ControlFreak, Salling Clicker is a two-part program that includes a controller agent module on the computer and a UI module on the phone. The controller agent is installed into the Other section of the Mac's System Preferences. You can configure the behavior of the remote control by clicking the Salling Clicker icon (see Figure 7-14).

Once you start the agent by clicking the Salling Clicker icon in System Preferences, it runs in the background and listens for Bluetooth connections from the phone. An icon in the system menu bar indicates the connection status (see Figure 7-15). You can also use the menu bar icon to quickly access Salling Clicker without going through the System Preferences window.

On the phone, you just need to install the Salling Clicker program. The program provides the phone UI for the remote control. You can initiate a connection from the Mac to the phone or from the phone to the Mac. Once the connection is established, it is kept active. From my experiments, it seems that computer-initiated connections are more reliable.

Now, with the software properly installed, let's explore the key functionality of the Salling Clicker program.

Control iTunes

The most commonly used feature of Salling Clicker is to control music playback via the iTunes software. From the phone UI, you can select the playlists and songs to play. The current song's name, album, artist information, and cover art image, if available, are displayed on the phone (see Figure 7-16).

You can also rate the songs right on the phone (see Figure 7-17). This is much more convenient than rating songs in the iTunes window, which requires you to interrupt your workflow and switch the currently active window to iTunes during playback. I have found myself rating songs much more frequently on the phone remote control.

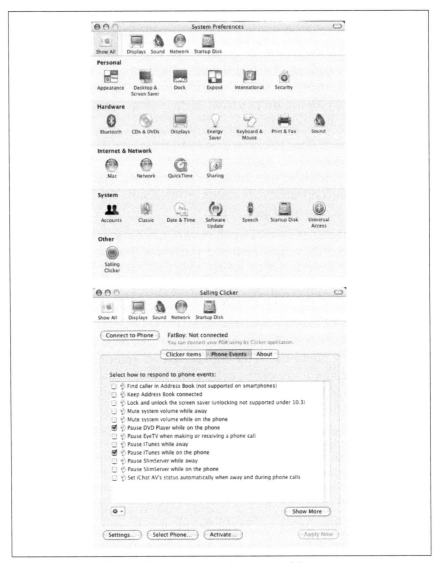

Figure 7-14. The Salling Clicker icon in the Other section of the Mac's System Preferences

With the Apple Airport Express device, you can stream audio from the iTunes library to any connected stereo speakers via the local WiFi network. Using the Salling Clicker remote control, you can make a Mac in the living room play music on a kitchen stereo while you are working in the kitchen.

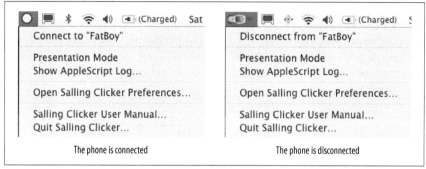

Figure 7-15. The system icon and menu for Salling Clicker

Figure 7-16. Play songs in iTunes

Figure 7-17. Rating a song in iTunes

Another cool Salling Clicker feature is that it pauses the iTunes playback when you pick up an incoming call, which minimizes interference.

Control iPhoto

Salling Clicker can launch iPhoto from the phone and then display any photo in the iPhoto library. The photo is displayed on the screens of both the Mac computer and the phone. It also starts photo slideshows. If you have a projector, this could make an excellent remote control for a photo demonstration. You can move to the next or previous photo using the phone navigation pad. Again, the slideshow pictures are shown on the phone screen as well, allowing you to view the show while you are away from the Mac (see Figure 7-18).

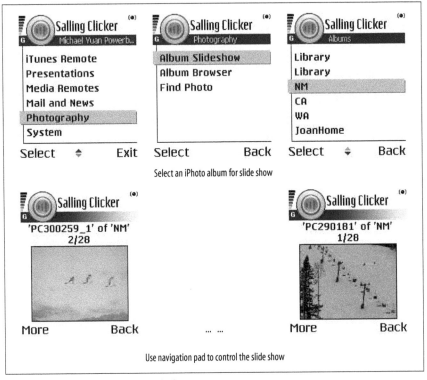

Figure 7-18. Running a photo slideshow

Control Video Players

Salling Clicker can control several popular video playback programs on the Mac. They include the following:

QuickTime player
    Plays generic media files, such as *.mov*, *.avi*, and *.mp3* files

DVDPlayer
    Plays DVD movies

VLC player

An open source video player that plays media files, VCDs, and DVDs

eyeTV player

A player for cable or satellite TV programs captured from a TV capture and conversion card

The computer can play video content on its own display screen, or output S-Video signals to show on a standard TV. Figure 7-19 shows the phone remote UI for QuickTime. It displays the current movie information and the player timer. You can use the navigation pad to play and pause the movie. You can also use the More option to open new movie files and toggle the full-screen mode.

Figure 7-19. Playing a movie

Control Presentations

Salling Clicker acts as a remote control for Apple's Keynote and Microsoft's PowerPoint presentation software. You can put the computer into presentation mode and use the navigation pad to move the slides forward and backward. Figure 7-20 shows that in PowerPoint's presentation mode, the phone screen displays the speaker notes and the title of the next slide. These are extremely powerful tools for effective presentations.

 Unlike ControlFreak, Salling Clicker does not display the slide image on the phone screen. But that is not a big issue, since you can also see the slide projected on the screen from anywhere in the room.

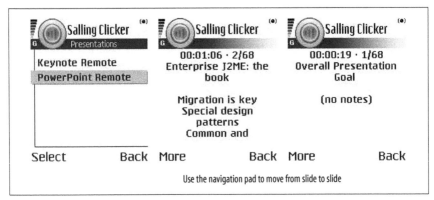

Figure 7-20. Controlling a PowerPoint presentation

Check Email and Read Blogs

You can launch the Mac Mail program and Ranchero Software's NetWire-News program (only the commercial version) from Salling Clicker. Figure 7-21 shows how to check and read email messages on the phone.

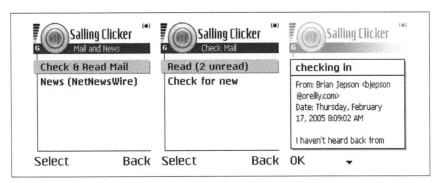

Figure 7-21. Checking and reading email messages

This allows you to read email and Really Simple Syndication (RSS) syndicated blogs while you are away from your computer. For instance, you can leave the computer on your desk and read email from your couch.

Salling Clicker controls only the Mac programs. The email and RSS contents are downloaded and rendered by the Mac before they are streamed to and displayed on the phone.

Control the Mac System

Using the Salling Clicker program, you can move and click the mouse on the Mac using the navigation pad. This function is similar to that in Control-Freak [Hack #41]. However, unlike ControlFreak, Salling Clicker does not display the computer screen on the phone. You can also adjust the system sound volume and sound balance from the mobile phone (see Figure 7-22).

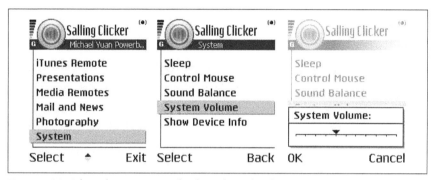

Figure 7-22. Adjust the system sound volume from the phone remote control

Control Any Mac Program

Now, you might be asking how Salling Clicker can support so many applications. Can it support any other applications that you want to control? The answer is probably yes.

A key feature of Salling Clicker is that it can invoke any AppleScript on the Mac. This allows it to potentially launch and control any program on the Mac. In fact, many third-party Salling Clicker scripts are available. They control everything from professional audio applications to telescopes. You can see a list of contributed scripts at *http://homepage.mac.com/jonassalling/ Shareware/Clicker/scripts20/index.html*.

> AppleScript is a scripting language on Mac OS 9 and OS X. It not only interacts with the operating system to launch and stop applications, but it also interacts with the applications themselves to provide specific services.

In the Salling Clicker control console in System Preferences, you can edit the controller categories and their associated AppleScripts. Figure 7-23 shows that in the Clicker Items tab, the left column lists the categories and the right column shows the available AppleScripts.

Figure 7-23. Controller categories and associated AppleScripts

You can click any AppleScript in the righthand column and open it in the system's AppleScript editor. The following listing shows the AppleScript to start and pause the DVD Player. It uses the SEC Helper application to display large gray alert messages on the Mac screen.

```
tell application "DVD Player"
    if dvd state is playing then
        pause dvd
        tell application "SEC Helper"
            show screen message "DVD Player" icon pause
        end tell
    else
        set viewer full screen to true
        activate
        play dvd
```

```
        tell application "SEC Helper"
            show screen message "DVD Player" icon play
        end tell
    end if
    return
end tell
```

By inspecting those source code files, you can learn the basics of Apple-Script quickly and get started writing a remote controller for your favorite Mac applications. To learn more about AppleScript, you can refer to the excellent book *AppleScript: The Missing Manual* (O'Reilly, 2005).

## CHAPTER EIGHT

Improve the User Interface
Hacks 43–49

The Nokia smartphone user interface (UI) is designed to make the best use of both the audio and the display elements on the phone. A well-designed UI increases user productivity on the phone and makes users more likely to use their phone's entertainment features. But different people have different ideas of what makes the best UI. Because a smartphone is a highly personal device, the smartphone UI must be highly customizable to match everyone's needs.

In this chapter, you will learn how to customize your phone's UI to your liking. I will also cover tips on how to obtain and create art elements for your custom UI.

HACK #43 Change Sound and Alert Settings on the Fly

Using profiles, you can change your phone's sound and alert settings with several quick clicks.

A *profile* is a collection of settings for the phone's ring tone, alert tones, sound volume, and vibrating alert. Using profiles, you can quickly change your phone's settings to adjust to your surroundings. For example, when you enter an important meeting, you can switch your phone into the Silent profile, which silences the ring tone and turns off all alerts, in one operation. On most Nokia devices, lightly pressing the Power button on the top of the handset brings up the available profiles. You can either use the navigation pad to select different profiles or simply tap the Power button multiple times to move the selection highlight down the list of profiles. Once you select a profile, hold down the power key or press OK (the left soft key) to activate the profile. This allows you to change profiles quickly with a minimal number of key presses.

In most professional environments, it is basic mobile phone etiquette to silence the ringer and to use vibrating alerts only. Profiles make this very easy to do.

Most phones have five built-in profiles: General, Silent, Meeting, Outdoor, and Pager. Their factory settings are listed in Table 8-1. On Series 60 devices, you can customize the settings for any of these profiles using the Profiles application in the Main menu. On Series 40 devices, you can customize the profiles via the Settings → Profiles → Personalise menu (see Figure 8-1). You can change the name of all the profiles except for the General profile.

Table 8-1. The factory-set profiles

Option	General	Silent	Meeting	Outdoor	Pager
Ringing type	Ring	Silent	Beep once	Ring	Silent
Ringing volume	Middle	n/a	Low	Loud	n/a
Message alert	On	Off	Beep twice	On	On
Chat alert	On	Off	Beep twice	On	On
Vibrating	Off	Off	On	On	On
Keypad tones	Middle	Off	Off	Loud	Middle
Warning tones	On	Off	Off	On	On

On older Series 40 devices (e.g., the Nokia 6800), Profiles is a top-level menu item rather than a submenu item under the Settings menu. Their functionality is the same.

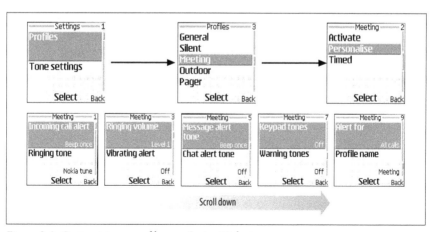

Figure 8-1. Customizing a profile on a Series 40 device

The Keypad tone is the confirmation sound you hear when any key is pressed. The Keypad tone can get very annoying when you are working with other people in the same room. But if you turn off the Keypad tone, you will not hear the touchtone when you dial. Keep this trade-off in mind when you change those settings!

Changing profile settings is a matter of personal preference. For instance, I really need a "vibrating only" profile. So, on my devices, I always customize the Pager profile into a Vibrate profile (see Figure 8-2). As a result, my phone vibrates for incoming calls, but does not make any sound, under any circumstances.

On Series 60 devices, changing the Ringing type to Silent turns off the sound for games and applications as well. On Series 40 devices, the game sound is separately configured in the Options menu for the games folder.

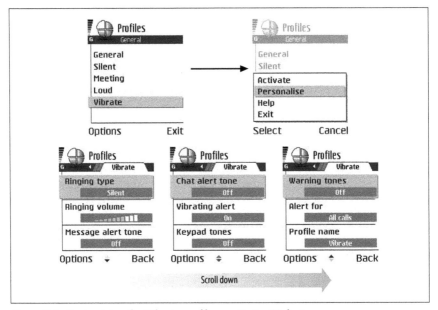

Figure 8-2. Customizing the Vibrate profile on a Series 60 device

You can personalize profiles in many other ways as well. For example, you can change the phone's ring tone for each profile. I change the name of the Outdoor profile to Loud, and the Nokia default ring tone to something that sounds louder. The Loud profile makes sure that I can always hear when the phone rings, even if I've got music playing.

Prioritize Your Callers

You have probably noticed that the profile personalization screen has an option called "Alert for." That is a very useful option, despite its cryptic name. You can use it to decide whose calls to take when a particular profile is activated. For instance, you can set up a special Meeting profile that rings only when your boss (or your spouse) calls and ignores all other incoming calls.

The Nokia Contacts list allows you to group your contacts into several different caller groups. Figure 8-3 shows the caller groups on a Series 60 device. You can click each group name to view and edit its membership or change its name. Caller groups are supported on Series 40 devices as well.

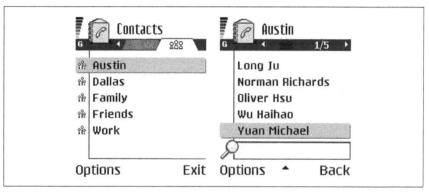

Figure 8-3. Caller groups and their memberships

For each profile, you can specify one or several caller groups in the "Alert for" option. The phone will alert you only when a member of one of these groups calls (or messages you), using the tones and volumes defined in the profile. Figure 8-4 shows how to configure the Meeting profile for caller groups. It only lets through calls from my family and from people in Austin, Texas.

Profiles give you a lot of control over various forms of feedback, accepting calls from people you know and ignoring (ringing silently) calls from those you don't know, and they can prove to be a valuable way to manage your time effectively. One of the most irritating aspects of an always-on society is that people *know* you're always on. Using profiles to set some limits frees you from making that decision on a per-call basis.

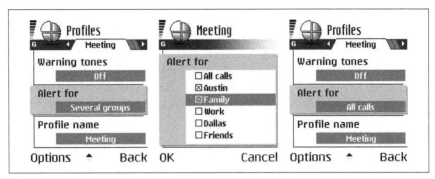

Figure 8-4. Configuring the Meeting profile to allow only certain callers to get through

 ## Automatically Switch Profiles

#44   Automatically change a profile based on time and location. This lets you make the phone adjust to its environment, without user intervention.

While changing the profile by hand is easy, many users still forget to do it. You often hear mobile phones ringing in meetings—someone forgot to switch to their Meeting profile. Wouldn't it be nice if the phone could automatically switch to different profiles based on the time of day, or even the phone's location? Well, you can do that with Nokia smartphones.

Change Profiles Based on Time

Although it is easy to change profiles using the Power key, people still often forget to do so. It is embarrassing to have your phone ring in the middle of a movie or a presentation, and the scramble to find the phone in your bag or briefcase is comical at best. Using a small Symbian program from Psiloc, called Extended Profiles Pro (available for purchase from *http://www.psiloc.com/ index.html?id=156*), you can change profiles at preset times on Series 60 devices. Figure 8-5 shows how to configure the Extended Profiles Pro program.

> Some non-Nokia devices support changing profiles based on calendar events. For instance, if the calendar knows you will be in a movie from 7:00 p.m. to 9:00 p.m., the phone can be automatically switched to the Meeting profile during that time. Unfortunately, this functionality is not available on Nokia phones.

Extended Profiles Pro does not support Series 40 devices. However, the Series 40 native profile manager allows you to set an expiration time for each profile when you activate it (see Figure 8-6).

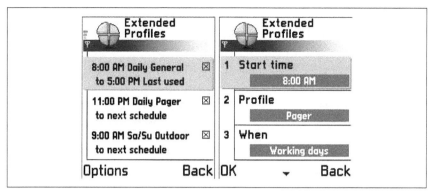

Figure 8-5. The Extended Profiles Pro program, which automatically switches profiles at preset times

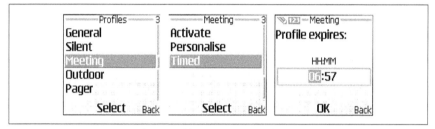

Figure 8-6. Setting time-out for a profile on a Series 40 device

Change Profiles Based on Location

Another Psiloc program for Series 60 devices, called miniGPS (*http://www.psiloc.com/index.html?id=155*), allows you to change profiles automatically depending on your (approximate) location. The Global System for Mobile communications (GSM) cellular network is serviced by a network of base stations. The area covered by each base station is known as a *cell*, which can be anywhere from 100 square meters in a crowded city to 3 square miles in the countryside. Using the area ID and the cell ID of the current base station, miniGPS determines the approximate position of the phone. For instance, if your office and home are more than 100 meters away from each other, they are likely to be serviced by base stations with different cell IDs.

Figure 8-7 shows that you can assign a location name to a unique combination of area ID and cell ID in miniGPS. Then, you can configure what profile to activate when your phone enters this location.

In the next several hacks, I'll discuss more UI customization tips related to profiles. So, read on.

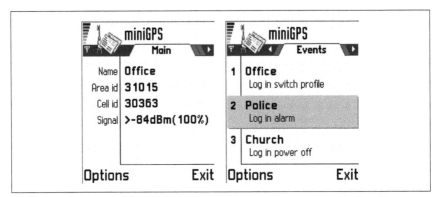

Figure 8-7. The miniGPS program, which automatically switches profiles based on your location

 ## Customize Ring Tones

Ring tones not only provide instant entertainment, but also can improve your productivity by identifying callers. Learn how to purchase ring tones or create your own for free.

Mobile phone ring tones are all the rage these days. According to an ARC study, the mobile phone ring tone business in 2003 was valued at $3.5 billion worldwide. That is 10% of the world music market and is many times bigger than the entire downloadable MP3 business in the PC world.

Use Ring Tones

A *ring tone* is the tone the phone sounds to alert you to incoming calls. You can associate a ring tone with a profile or with callers.

Set a ring tone for a profile. On most Nokia devices, you can specify a different ring tone for each profile. On a Series 60 device, you can just open the Profiles application and select the Options → Personalise menu. When you open the "Ringing tone" item, you are presented with a list of all audio files stored in the device (i.e., in the Gallery). You can choose any of these audio files as the default ring tone for this profile (see Figure 8-8).

Your phone should have come with quite a few ring tones preinstalled. To hear a ring tone, scroll to it in the list and pause for a few seconds.

On a Series 40 device, you can select a profile from the Settings → Profiles menu and then select Options → Personalise to change ring tones. If you want to select a ring tone from an audio file in the phone's default Gallery folder, choose the Open Gallery item in the list of available ring tones (see Figure 8-9).

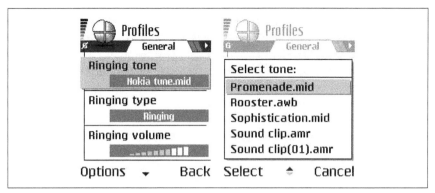

Figure 8-8. Setting a ring tone for a profile on a Series 60 device

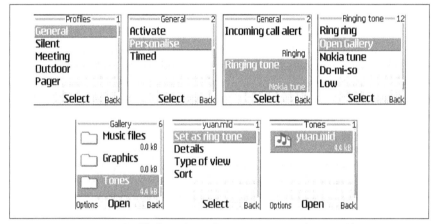

Figure 8-9. Using an audio file from the Gallery as a ring tone for a profile on a Series 40 device

On Series 40 devices, you can use the Settings → Tone settings menu to customize ring tones and alert settings for a currently active profile. On some devices, you can also navigate directly to an audio file in the Gallery and then set it as the ring tone for the current profile via the Options menu.

Set a ring tone for a caller or caller group. On Nokia Series 60 phones, it is also possible to assign a different ring tone for each contact or each caller group. On Nokia Series 40 phones, you can specify up to five ring tones to five different caller groups, but you cannot specify ring tones for individual contacts. Those per-contact or per-caller group ring tones allow you to differentiate callers by the sound of the incoming call. Please refer to "Put a Face or Tune to the Caller" [Hack #27] for more on this subject.

Create Free Ring Tones

You can purchase and download ring tones for your phone from many web sites, including your wireless operator's content portal site. Many leading musicians now sell ring tones for their new songs as both a revenue stream and an effective promotional vehicle. Part of the commercial success of the mobile phone ring tone business lies in the fact that, unlike MP3 music on PCs, ring tones are hard to bootleg and share. But advanced mobile phone hackers can create and use ring tones for free, from music they already own.

Use MIDI songs. MIDI is the acronym for Musical Instrument Digital Interface. It is by far the most popular ring tone format. A MIDI file contains the musical notes of a song instead of the actual audio recording. Therefore, MIDI files can be very small, which makes them ideal for mobile phone ring tones.

> MIDI songs are music only. MIDI cannot reproduce the human voice or other sounds that cannot be represented by musical notes.

The musical notes in a MIDI file are divided into several parallel channels. When a MIDI player plays the file, it simulates a different musical instrument for each channel and plays all the channels simultaneously. That produces the effect of a whole band playing the song together. Obviously, the more channels you have in a MIDI file, the better it sounds.

Many free MIDI files for popular songs are available on the Internet. However, most MIDI files you download cannot be used as ring tones because most free MIDI files have more than 256 channels. A Nokia mobile phone can play only 4–24 notes at a time due to hardware limitations. The Nokia Sound Converter program in the Nokia PC Suite **[Hack #15]** allows you to reduce the number of channels on any MIDI file and create Scalable Polyphony MIDI (SP-MIDI) files that are suitable for a selected Nokia device (see Figure 8-10). Using this tool, you can define channel priorities for the conversion, which specifies which channels to play and which channels to omit.

Compose your own MIDI songs. If you are talented in music composition, you can use the Composer program included in some Nokia Series 60 phones (e.g., the N-Gage QD) to compose MIDI tones on the fly. The Composer is also very useful if you have access to the musical scores of your favorite songs but not the actual MIDI files. Figure 8-11 shows the Composer in action.

Convert MP3s to ring tones. Compared with MIDI music, MP3 songs make much more impressive ring tones. MP3 files can incorporate human voices

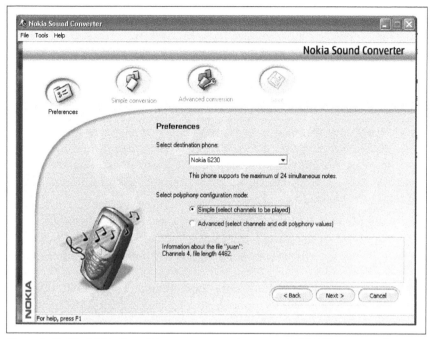

Figure 8-10. The Nokia Sound Converter converting general MIDI files to ones that can be played on Nokia devices

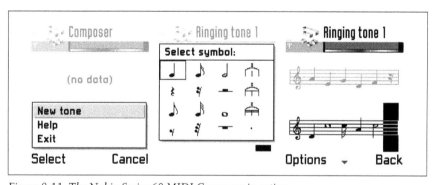

Figure 8-11. The Nokia Series 60 MIDI Composer in action

and other sound effects. Most important of all, you probably already have your favorite music collection in MP3 format.

Unfortunately, most Nokia phones cannot use MP3 files directly as ring tones. The hack here is to first convert MP3 music to Audio/Modem Riser (AMR) or WAV audio files, which Nokia devices do support. The AMR and WAV files can hold the same type of sound as MP3 files, but the former are less compressed.

You have to be careful with the size of the AMR and WAV files. A full-length song in *.wav* format can be more than 50 MB. That is much larger than the memory space of most Nokia phones. I recommend you convert only a 5 to 10-second clip of your favorite MP3 song into an AMR/WAV ring tone.

The Nokia Multimedia Converter is a standalone application outside of the Nokia PC Suite. It converts a variety of PC audio/video formats to mobile phone formats. Also, it supports MP3-to-AMR conversion. You can download the Nokia Multimedia Converter program for free from Forum Nokia (*http://www.forum.nokia.com/*).

Also, many MP3-to-WAV file converter programs are available on the Internet as shareware or freeware. For instance, the very popular and free MP3 player, Winamp (*http://www.winamp.com/*), can export MP3s to *.wav* files.

Record your own ring tones. Many Nokia phones come with a built-in sound recorder. With it, you can record a short voice clip into an AMR file in the Gallery and then use the AMR file as a ring tone. For example, you can record your spouse saying "It's me! Pick up the phone," and assign it as the ring tone for him. Using voice clips as ring tones is a great way to personalize your mobile phone experience.

Use Ring Back Tones

I have said enough about ring tones, which you hear when people call you. The latest twist in the ring tone fad is to use *ring back tones*, which are also known as *caller tones*. They are the tones your callers hear when your phone rings, instead of the "ring, ring" they usually hear. This is a great way to entertain your friends and show your taste to your callers. To use this service, your wireless operator must support it. You can purchase ring back tones from your operator's portal site. However, if you change your operator later, all your ring back tones will be lost.

Customize the Idle Screen
Add eye candy to your phone via wallpaper, an operator logo, and custom fonts. Also, enhance the UI for the elderly and vision impaired.

The idle screen (a.k.a. the home screen) is the "face" of your Nokia mobile phone. It is one of the defining features of the phone's exterior style, especially for the very popular "candy bar" phones. You see this screen every time you pick up your phone. It is definitely worth spending some time to customize the idle screen to your taste.

The customizable visual elements on the idle screen include the wallpaper, the operator logo, and the font. I'll discuss all of them in this hack.

Set an Image as Wallpaper

Wallpaper is the background image displayed on the idle screen. On most Series 60 devices, the wallpaper occupies only the central part of the screen. On most Series 40 devices, the wallpaper fills the entire screen (see Figure 8-12).

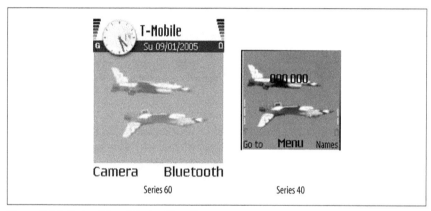

Figure 8-12. Series 60 and Series 40 wallpapers

You can set any image in your phone's Gallery as wallpaper. On most new Nokia phones, you can simply open the image in the Gallery and select the "Set as wallpaper" item from the Options menu (see Figure 8-13). If you have an older Series 60 phone (e.g., a Nokia 3650), you should go to Settings and customize the wallpaper for the standby mode (see Figure 8-14). On a Series 40 device, you can also use the Settings → Display Settings → Wallpaper menu item to set your wallpaper. The phone automatically resizes or crops the image as needed to fit the wallpaper area on the screen.

> When you set new wallpaper, the old wallpaper is discarded without warning. If you want to revert to the old wallpaper, you need to keep a copy of it in the Gallery. I recommend that you create a "wallpaper" folder in your Gallery so that you can switch back and forth between wallpapers whenever you want.

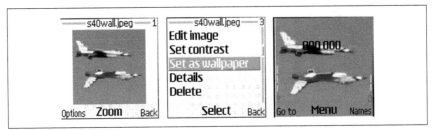

Figure 8-13. Setting wallpaper from an image in the Gallery

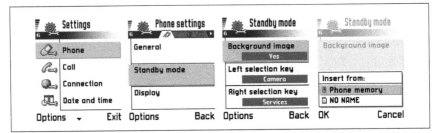

Figure 8-14. Setting wallpaper on a Nokia 3650 device (older Series 60 device)

The Extended Profiles Pro program from Psiloc [Hack #31] supports associating wallpapers with profiles on Series 60 devices. This way, when you change the profile, you change not only how the phone sounds but also how it looks. When you use Extended Profiles Pro together with the miniGPS program [Hack #44], you can switch wallpapers based on your current location.

Create Free Wallpaper

Many commercial web sites, including most wireless operators' portal sites, sell downloadable mobile phone wallpaper for $1 to $3 apiece. But truth be told, it is very easy to create professional-looking wallpaper yourself, for free. The basic idea is to first select an image you like, resize or crop it to fit the phone screen, and then convert it to a supported image format before transferring it to the phone.

If you own a digital camera, you probably have many everyday or vacation pictures you can use as phone wallpaper. And if you've got a camera phone, you probably have plenty of wallpaper candidates sitting right in your Gallery. But if you are not the photographer type, plenty of free images are available on the Internet. A good place to start is Google's image search, at *http://images.google.com/*. I discuss more tips on capturing fancy images later in this section.

As I mentioned before, when you set an arbitrary image as wallpaper, the phone automatically resizes or crops it. But it is much better if you process the image yourself, since this saves memory space and bandwidth. It also eliminates the uncertainty associated with the results of the phone's image processing.

To turn an arbitrary image into a piece of wallpaper, you need an image editor program. I recommend GIMP (*http://www.gimp.org/*), which runs on Windows, Linux, and Mac OS X computers. You can also use the Paint program that comes with every Windows PC—but the resizing quality of Paint is poor. You should resize or crop the image to 172×143 pixels for Nokia Series 60 phones and to 128×128 for Nokia Series 40 phones. Nokia devices support *.bmp*, *.jpg*, *.gif*, and *.png* image formats. If your image is in a different format, you should save it to one of the supported formats.

Since the wallpaper overlaps with the information displayed on the phone screen, I recommend that you use images that are light in color as wallpapers. If you have a piece of wallpaper with large areas of dark colors, you probably want to reduce the contrast and increase the brightness to make it look "washed out."

Figuring out and remembering the exact wallpaper image size for your specific phone is kind of a hassle. But with the Nokia PC Suite, you do not have to do that! The Image Converter program in the Nokia PC Suite helps you convert any image to a size and format suitable for the wallpaper on your phone (see Figures 8-15 and 8-16).

Capture a movie scene. A very popular type of mobile phone wallpaper is a movie snapshot. You can capture scenes in a movie by playing the movie (or the trailer) on your PC and then capturing the screen. On a Windows PC, use the Print Screen key to capture the entire screen into the clipboard. Then, in an image editor program (GIMP is recommended, but Paint works too), you can paste the captured screen, crop the movie scene, resize it, and then save it to a supported format. On a Mac, you can use the Shift-Command-4 key (the Command key is the one with the Apple logo on it) to capture a rectangular area of the screen into a PDF file (or PNG file in Mac OS X 10.4) and then use an image editor program to convert it to a size and format suitable for mobile phones.

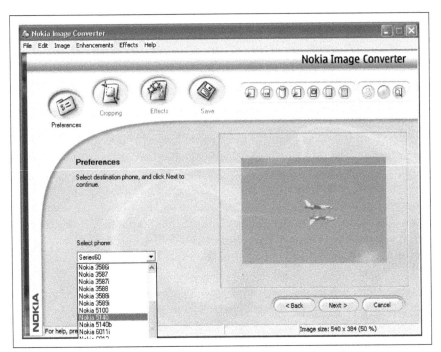

Figure 8-15. Choosing a target phone for creating wallpaper in the Nokia Image Converter

> Some movie players disable the system screen capture key.
> VLC (*http://www.videolan.org/vlc/*) is a free DVD and VCD
> player that supports screen capture.

Capture an image with a camera phone. If you have a camera phone, you can make some great wallpaper using your own images. This is a great use of your phone camera, and it adds a personal touch to your phone. For instance, you can use your loved one's pictures as more up-to-date substitutes for wallet pictures. When you are at a movie theater, you can take a picture of your favorite movie poster that is not yet available for downloading on the Internet. When you see a great magazine or newspaper photo depicting a current event or the most recent fashion, you can quickly snap it with your phone camera.

> When I see a great image on the Internet or a good DVD
> movie scene, I often take a picture directly from the PC
> screen with my phone's camera. It is much quicker than
> downloading or capturing the image on the PC, manipulat-
> ing it, and then copying it to the phone.

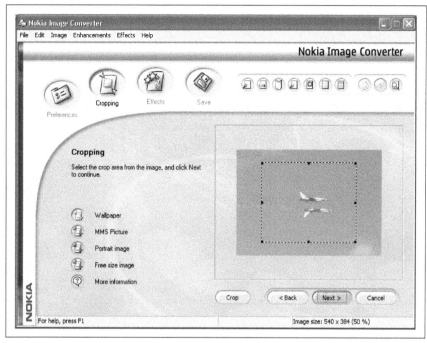

Figure 8-16. Cropping the wallpaper for the selected phone in the Nokia Image Converter

All in all, wallpapers are simple and effective ways of customizing your phone. If you have a camera phone, you have no excuse not to decorate it with personal pictures!

Change the Operator Logo

The *operator logo* is the word appearing at the top of the idle screen. On Series 60 phones, you can change the operator logo to your own logo using the freeware FExplorer tool [Hack #20]. On Series 40 devices, you can change the operator logo with a custom theme [Hack #49]. In this section, I'll discuss how to change the operator logo for a Series 60 phone.

The first step is to create an image of your own logo. The image must be 97×25 pixels in size. You can use any image editor to create it and then save it in one of the image formats supported by the phone.

After the logo image is copied to the phone, highlight it in the FExplorer file-system browser. Then use the Options → File → Set as operator logo menu to set it as the operator logo. The phone reboots and the new logo appears (see Figure 8-17).

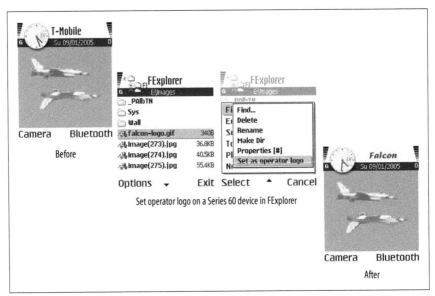

Figure 8-17. Setting the operator logo on a Series 60 device with FExplorer

To eliminate the custom operator logo, use FExplorer to navigate to the C:\ system\Apps\phone\oplogo directory and delete the file in it.

Change the Font

HACK #47

Change the display font on your Series 60 smartphone!

Mobile phone gurus at the Zedge Forum (*http://www.zedge.no/*) collected a set of Symbian programs that enable you to change the display font on Nokia 6600 and 7610 devices. Here's how the process works:

1. Download the zip package of Symbian programs from *http://www.green-ocean.net/fonts_for_6600_7610.zip* and unzip it into a folder.

2. Install and run the *FontRemover.sis* application on your device. You will be asked to remove the existing font and then reboot your device. *Do not* reboot at this moment.

3. Install and run one of the font programs on your device. You can have only one font at a time on any device! Make sure you install it in the phone's main memory—*not* in the MultiMediaCard (MMC) card. The *tahoma.sis*, *modern.sis*, and *mssans.sis* programs install the Tahoma, Modern, and MS Sans fonts, respectively.

4. Reboot the phone and enjoy the new fonts!

This technique is known to work on Nokia 6600 firmware v3, v4, and v5. It is not widely tested on other Series 60 firmware. Use it at your own risk! If you have a problem, you might have to do a hard reset of your phone to return it to the factory settings [Hack #23].

This hack not only changes the fonts on the idle screen, but it also changes all the menu and label fonts throughout the device. Figure 8-18 shows the Modern and Tahoma fonts installed on a Nokia 6600 smartphone.

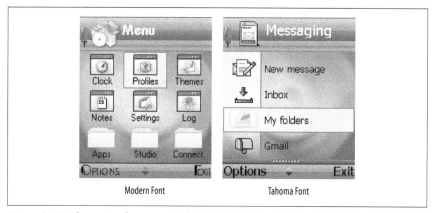

Figure 8-18. Alternative fonts on a Nokia Series 60 smartphone

It is important to note that this hack has not been tested on all Nokia Series 60 smartphone firmware versions. So, use it at your own risk! According to a forum posting from slurpy69 at Zedge, dated January 23, 2005, the following phone models and firmware versions are known to work with this hack:

- Nokia 6600 v3.42.1, v4.09.1, and v5.27.0
- Nokia 6630 v2.39
- Nokia 7610 v4.0437.4
- Nokia N-Gage v4.03 Nem-4

This hack does not work on Nokia 7610 firmware v4.0424.4.

You can determine the firmware version of your device by entering the code *#0000# in the idle screen.

HACK
#48

A User Interface for the Vision Impaired

Tools allow the vision impaired and senior citizens to enjoy their phones, despite the phones' small screens and tiny fonts!

To display more content on their limited screen, mobile phones use small fonts. That is very inconvenient for elderly users and the vision impaired. Several Symbian applications are available to help make the Nokia Series 60 smartphones easier to use for those users.

One category of applications is *screen readers*. They have text-to-speech engines to read out everything that is currently displayed on the screen via the phone's speaker. Of course, those screen readers run in the background to avoid interference with regular phone use and with other applications. Popular screen reader programs include:

- The SpeechPAK TALKs program from ScanSoft (*http://www.scansoft.com/speechpak/talks/*)

- The Mobile Speak program from Code Factory (*http://www.codefactory.es/*)

> Screen readers not only help the vision impaired to use their phones, but also allow hands-free access to email and Short Message Service (SMS) messages on the phone. They also work great with Bluetooth car kits.

Another type of program actually magnifies part of the screen to make the text easier to see. For instance, the Mobile Magnifier program, also from Code Factory, runs in the background and magnifies the currently highlighted item on the screen so that it is easier to see (see Figure 8-19).

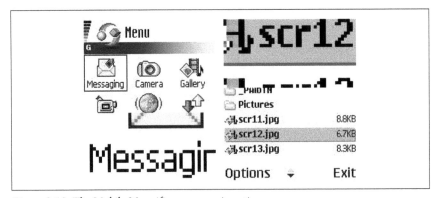

Figure 8-19. The Mobile Magnifier program in action

Develop and Use Themes

Using themes, you can simultaneously change many elements of the phone UI, especially on Series 60 phones.

Themes allow you to fully customize many visual and audio elements on your phone UI beyond the wallpaper and the ring tone. Themes are supported only on relatively new Nokia phone models:

- For Nokia Series 60 phones, only the ones that were released at the same time as or after the Nokia 6600 support themes. Examples are the Nokia 6600, 6620, 7610, and 6630 devices.

- For Nokia Series 40 phones, only the ones that were released at the same time as or after the Nokia 3220 support themes. Examples are the Nokia 3220, 6020, 6170, 7260, 7270, and 7280 devices.

Figures 8-20 and 8-21 show a Series 60 theme and a Series 40 theme, respectively. Both of them are developed by MANGOobjects and are available for purchase at *http://www.mangothemes.com/*. I will discuss how to install these themes later in this hack.

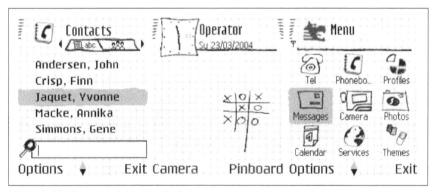

Figure 8-20. An example Series 60 theme

Figure 8-21. An example Series 40 theme

On a Series 60 device, you can customize all UI elements in the following list via themes:

- Elements of the idle screen, including the wallpaper
- Menu background
- UI areas, including status, list, and column areas
- Icons, including application, list, and note icons
- Highlights on list, grid, and input areas
- Pop-up windows
- System indicators, including volume, tab, signal strength, battery charge, navigation, and wait/progress bar indicators
- Color scheme
- Sounds, including ring tone and message alert
- Screensavers

On a Series 40 device, however, your customization choices are much more limited. You can change only the ring tone, wallpaper, operator logo, screensaver, and color theme. Nokia is actively working on expanding theme support for future Series 40 devices.

> On a Nokia Series 60 smartphone, you can associate a theme with any screensaver installed on the device. Mirkocrocop posted a tutorial on NokiaFree forums on how to quickly make your own screensavers from a series of images. It is verified to work on Nokia 6630 smartphones. Visit *http://nokiafree.org/forums/showthread.php?p=364259* to read the tutorial. Darla Mack expanded the method and discussed how to copy screensavers from one device to another on her blog: *http://darlamack.blogs.com/darlamack/2005/05/new_way_to_make.html*. Check them out!

Create a Theme

Nokia offers two free tools that content developers can use to create themes—the Nokia Series 40 Theme Studio and the Nokia Series 60 Theme Studio (for Symbian OS), both of which are available for download from the Tools and SDKs section of the Forum Nokia web site (*http://www.forum.nokia.com/tools*).

Using those tools, you can easily design and develop personal themes for your Nokia phones. You can share your themes with friends on the Internet, or even sell them to the public for profit.

Create themes for Series 60 devices. The Series 60 Theme Studio (for Symbian OS) creates themes for Series 60 devices. The Series 60 Theme Studio is downloaded as a zip file containing an installation *.exe* file, a release notes text file, and the Series 60 Theme Studio Artist's Guide in PDF format, which provides a comprehensive guide to using the tool. You have to register for a free user account with Forum Nokia first and log into the web site before you can download the zip file. During installation, which is straightforward, you can specify an external image and sound editor for editing theme elements, although these can also be defined after the tool is installed.

When you launch it, the Series 60 Theme Studio (for Symbian OS) provides a menu-like interface for defining the various visual and audio components of a Series 60 theme. You can build a Series 60 theme to provide custom images for almost every component of the Series 60 UI, even to the extent of defining custom icons for third-party applications. It is not necessary for every theme to contain all the customizable elements, and when a new theme is created, the tool provides options to define which elements will be customized. These customized elements can also be based on the contents of an existing theme.

The tool supports the definition of all theme components using BMP image files, except for screensavers (which are created as a Symbian OS DLL and are included in the theme package) and the ring tone and message alert (which can be defined as WAV, MP3, or device-specific-supported ring tone formats). Once the theme's components have been defined, the Series 60 Theme Studio (for Symbian OS) allows you to simulate all of the device's main screens, screen types, and certain applications, such as the Calendar and Contacts, as shown in Figure 8-22.

Once you've created a theme, the tool lets you save it as a Symbian installation file (*.sis* file), with options for digital signing and Digital Rights Management (DRM). Once these steps have been completed, the theme is ready for delivery to your phone, friends, or customers.

Create themes for Series 40 devices. The Nokia Series 40 Theme Studio is downloaded as a zip file that contains a single *.exe* installation file. Installation is straightforward but does require a license key, which any registered member can obtain free of charge from Forum Nokia. You can request the key in advance or during the installation process.

When started, the Nokia Series 40 Theme Studio provides a simple interface for defining the components of a Series 40 theme: the color scheme, ring tone, wallpaper, and screensaver for UIs of 128×128 pixels and 128×160 pixels.

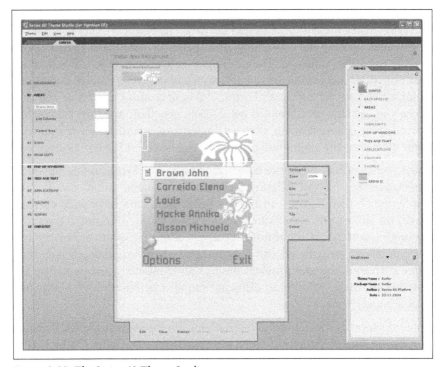

Figure 8-22. The Series 60 Theme Studio

The tool supports the definition of wallpaper and screensavers using files in JPEG, GIF, PNG, BMP, or Wireless Bitmap (WBMP) image formats, and ring tones as AMR or MIDI format files. Once the theme's components have been defined, the Nokia Series 40 Theme Studio allows you to simulate all of the device's main screens and screen types, such as the main and idle screens or lists, notes, and find screen types, as shown in Figure 8-23.

Once you've created a theme, the tool saves the content as a Series 40 Theme package file (*.nth*) ready for delivery to the phone.

Install a Theme

The Series 40 and Series 60 themes are packaged in *.nth* and *.sis* files, respectively. You can either create your own theme packages using the tools described in this hack or purchase and download commercial theme packages from a web site. You can install the theme packages via the following means:

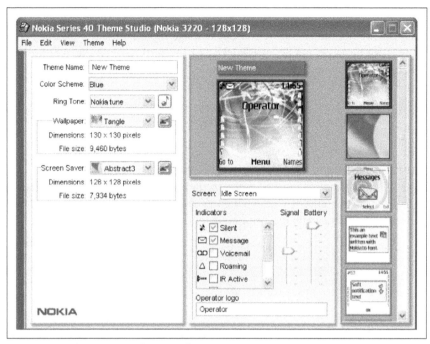

Figure 8-23. The Series 40 Theme Studio

- You can download and install the themes directly Over The Air (OTA) via the phone's Services browser. It is, however, necessary to set the correct MIME types on a download server **[Hack #53]**: *application/vnd. symbian.install* for Series 60 theme **.sis* files; and *application/vnd.noks40theme* for Series 40 theme **.nth* files.

- You can transfer the theme package files to the device via Bluetooth and IR. Once you open the message containing the file, the phone prompts you to install it.

- You can transfer a theme package to a phone by sending it as an MMS message attachment to the phone. You will be prompted to install it when the message is opened. The only limitation that could apply is if there is a size restriction on the delivery mechanism, such as the 100KB limit often placed on MMS messages. In this case, you will need to consider the size limitation when you create the theme.

You can install multiple themes on a device and switch between them at any time.

Manage Themes

Once the themes are installed, you can manage them on the phone using the Themes application on a Series 60 device or the Settings menu on a Series 40 device.

 If your phone supports themes, it probably shipped with more than one theme preinstalled.

The theme management application allows you to see a list of installed themes, and then preview and activate them. You can also do some basic customization for any of the installed themes on the phone. For instance, you can change the wallpaper, ring tone, and color scheme.

—Courtesy of Forum Nokia

The Mobile Web
Hacks 50–57

The World Wide Web is a killer application for PCs. It was widely expected that the Web would become a killer application on connected smart mobile devices as well. After all, the ability to access information, such as news, weather, movie tickets, and Yellow Pages listings, at any time from anywhere is very appealing. However, accessing the Web from a small mobile phone is not as straightforward as accessing the Web from a desktop PC. You have to watch out for several things. Tools such as mobile portals and search engines can significantly improve the overall web-browsing experience for mobile phone users. In the first half of this chapter, you will learn how to make the best use of resources on the Web. You will also learn how to develop and publish your own web sites for mobile phones.

In addition to traditional web sites, weblogs (a.k.a. blogs) have emerged as an important form of web publishing in recent years. Blog content typically includes a personal diary, commentary, and links to other web sites, all organized in reverse chronological order. Images are also often included in blog posts. As smartphones become ubiquitous in our society, using the phone to read and update blogs from anywhere at any time gives both blog readers and authors new degrees of freedom. In the second half of this chapter, you will learn how to take advantage of your Nokia smartphone to read and post to blogs.

According to Wikipedia, the definition of a *blog* is as follows: "A weblog, a web log, or simply a blog is a web application that contains periodic, reverse chronologically ordered posts on a common web page. Such a web site would typically be accessible to any Internet user."

Browse the Web

Mobile browsers allow users to interact with WML and HTML web sites from their phones. In this hack, I'll cover the basics of the mobile Web and look at several popular browsers available on Nokia phones.

The web browser is one of the first applications that became available on smart mobile phones. Today, almost all Nokia mobile phones on the market have integrated web browsers.

However, after years of development, browsing the Web on a smartphone is still difficult for many users. The small screen size, slow and unreliable networks, limited content, and expensive data plans resulted in poor customer experiences and hindered the adoption of the mobile Web. In the past several years, much progress has been made on mobile browsers to improve users' experiences. In this hack, I'll discuss the landscape of the mobile Web, and how to leverage new browser technologies on your Nokia phone to browse the Web like a pro!

Why So Many Mobile Browsers?

On the desktop, the web browser is, at least at its core, a relatively simple application. It parses and renders Internet pages written in Hyper Text Markup Language (HTML). On a mobile phone, however, web browsers need to handle web pages written in a variety of markup languages. Web browsers on Nokia devices support three types of markup languages: HTML, WML (Wireless Markup Language), and XHTML MP (XML HTML Mobile Profile). A more detailed description of these languages is available in "Create a Mobile Web Site" [Hack #53]. In this hack, I'll introduce them from the phone user's perspective.

HTML
> HTML is the Internet standard for authoring web pages. HTML browsers have access to the widest range of online content. However, HTML is also considered too heavyweight for mobile phones. Most HTML pages are not designed for small screens and require special browser rendering tricks to make them fit into the phone screen.

WML
> WML is a simple web page–authoring language specially designed for phones. It supports a limited set of presentation elements suitable for most mobile phones. Hence, WML pages are small in size and quick to render. However, WML's presentation capability is inadequate for high-end devices, such as Nokia Series 60 devices. A large number of WML web sites exist for mobile phones.

XHTML MP

XHTML MP is a subset of XHTML, which is the well-formed XML version of HTML. The benefits of well-formed XML documents are that they are easy to parse and are less error prone. The presentation capability of XHTML MP is between that of HTML and WML. As XHTML is being adopted to replace HTML, XHTML MP browsers could access content authored for both phones and desktop computers.

Read on to find out how popular web browsers on Nokia phones support these content types.

The Nokia Browser

Nokia phones ship with a default web browser, called Services. It is available from the Main menu on both Series 40 and Series 60 devices. This is a WML and XHTML MP browser. For older Series 40 devices, the Services browser might support only WML. Refer to the product specification and manual on the Nokia web site to find out exactly which markup languages your browser supports. The key benefit of using the built-in Services browser is that it is tightly integrated with the phone. For example, you can download and upload files from the phone's local storage **[Hack #53]**, install Java applications Over The Air (OTA) **[Hack #16]**, and make phone calls directly from a web page via special links **[Hack #54]**.

On a Series 60 device, the browser displays the Bookmarks page when you launch it. It is loaded with default bookmarks from Nokia and your wireless operator. You can delete any of them, or add your own, via the Options menu. You also can organize the bookmarks into folders.

You have to click a bookmark to be connected to the data network. Once you are connected, the Options menu offers several more choices, such as manually loading from arbitrary URLs or disconnecting from the network. Figure 9-1 shows how to use the Services browser to connect to Yahoo!'s mobile home page.

> You can enter a new URL in the Series 60 Services browser in two ways. You can add a new bookmark and then click it. Or you can load any existing bookmark to connect to the network, and then use the Options → Go to menu to load a new URL.

On a Series 40 device, the Services browser displays a set of menus upon startup. From there, you can manage the bookmarks, adjust the settings, or load arbitrary URLs. Figure 9-2 shows Yahoo!'s WML page loaded in the Series 40 Services browser.

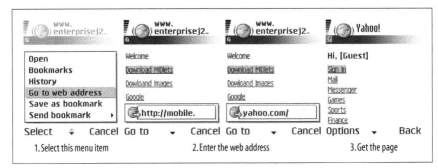

Figure 9-1. Yahoo!'s mobile home page displayed on a Series 60 device's Services browser

Figure 9-2. Yahoo!'s mobile home page displayed on a Series 40 device's Services browser

Besides WML and XHTML MP pages, the Services browser in newer models (e.g., the Nokia 6600, Nokia 6230, and later devices) can also display HTML pages. Figure 9-3 shows how the O'Reilly home page looks in a Nokia 6600 device's Services browser. Notice that the images on the page are automatically resized to fit the screen so that the user does not have to scroll horizontally to view the content. However, the toolbars and the table-based layout of the page do not render well in the small screen. It is also very slow to load, especially on a Series 40 device where the CPU power is inadequate to parse the complex HTML document. In the rest of this hack, I'll discuss better ways to browse HTML pages.

> The Services browser on Series 40 devices seems to render HTML content, especially tables, better than the Services browser on Series 60 devices, despite the larger screen size available with the latter.

The Opera Browser

The Opera browser is the leading HTML browser for Nokia Series 60 devices. Besides basic HTML markup content, Opera also supports Java-Script embedded in the pages. In many markets, Nokia 6600 phones are sold with Opera preinstalled. If your handset does not include Opera, you

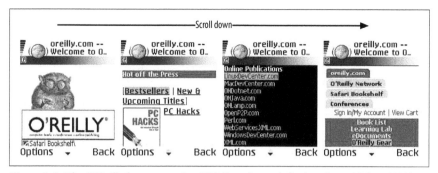

Figure 9-3. The O'Reilly home page (an HTML web page) displayed in a Nokia 6600's Services browser

can purchase a copy from *http://www.opera.com/* and install it yourself [Hack #17]. The Opera browser requires a relatively large amount of free memory to run. On an older device, such as the Nokia 3650, you must exit most background programs before you can start Opera. You can press and hold the menu key to see a list of background programs and use the c key to exit them one by one.

> If the Opera web browser's memory footprint is too large for your phone, you can try Doris—another HTML browser for Nokia Series 60 phones. Doris is much lighter than Opera, and it runs on the early Nokia 7650 phones. However, compared with Opera, Doris has a limited set of features. You can download and purchase Doris from *http://www. anygraaf.fi/browser/indexe.htm*.

Opera develops its web browser for many operating systems, including Windows, Mac OS X, Linux, Solaris, FreeBSD, OS/2, QNX, Symbian, and Windows Mobile. Since the Opera browser for Nokia Series 60 devices is based on the same core code as the award-winning Opera browser for PC and Mac computers, it can parse and render a great majority of HTML pages on the Internet, including sloppily formatted ones. Opera supports common browser features such as HTTP cookies, a cache for pages and images, JavaScript, HTTP proxy, HTTPS, and HTTP Basic authentication. It also supports full-screen browsing to make better use of the extremely limited screen real estate on a mobile phone (see Figure 9-4). In full-screen mode, the softkey labels at the bottom of the page are also hidden. But the left soft key is still mapped to the Options menu and the right soft key is mapped to Back or Stop, depending on whether the browser is currently loading a page.

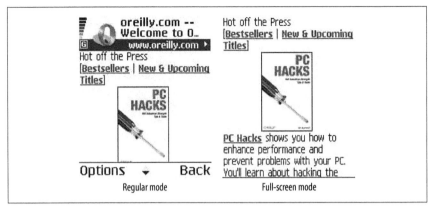

Figure 9-4. The Opera browser in regular and full-screen modes

 Opera defines a set of hotkeys to allow quick access to features without going through the menu. For example, pressing 1 pops up a box for entering a new URL to load; 2 opens
the bookmark page; 8 brings up the browser settings page; 9
pops up a box for entering a search phrase for a Google
search; and * toggles full-screen mode on and off.

A key innovation in mobile versions of the Opera browser is its content layout management. When you turn on Small Screen Rendering (SSR) mode in
the Options → Display menu (or use the # hotkey to toggle it on and off),
Opera will try to resize images and rearrange elements in the HTML page to
make it fit into the width of the screen. You can still use the vertical scroll
key to scroll the page. The pages in SSR mode will not necessarily follow the
original visual design of the web site. But they are more useable and more
effective on a small mobile phone screen. Figure 9-5 shows the layout of the
O'Reilly home page on a desktop browser, and Figure 9-6 shows its SSR
mode layout on the Opera browser on a Nokia Series 60 phone. Notice that
the toolbar is rendered in a compact format that is effective for small
screens.

 If you're a web developer, you might want to check how
your page will look in Opera's SSR mode. Simply select the
View → Small screen menu option on a PC or Mac version of
the Opera browser.

If you want to browse the page as its designers intended, you can turn off
the small screen mode. In that case, the page is probably much wider than
the screen. So, you need to scroll both horizontally and vertically to see the

Figure 9-5. The O'Reilly home page displayed in a desktop browser for comparison purposes

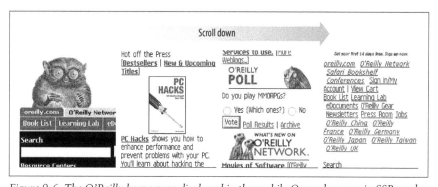

Figure 9-6. The O'Reilly home page displayed in the mobile Opera browser in SSR mode

entire page. Figure 9-7 shows the O'Reilly home page in this mode. The figure is a 5×4 grid of screens that simulate scrolling in both the horizontal and vertical directions.

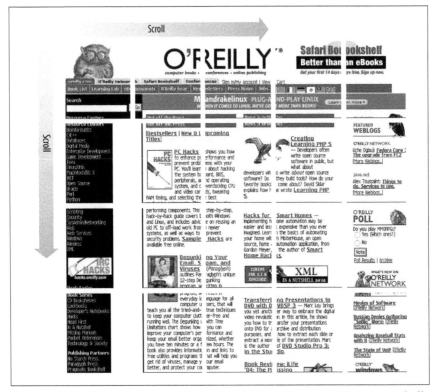

Figure 9-7. The O'Reilly home page displayed in the Opera browser with SSR turned off

The Opera Browser with Proxy

The standalone Opera browser is powerful and feature rich. But it does not solve the fundamental issue with wireless networks that hinders the adoption of the mobile Web: the mobile data network is slow and often expensive. For graphics-intensive web pages, the Opera browser needs to first open multiple HTTP connections to download all the images, and then resize them to fit the device screen if SSR is used. That is a slow and bandwidth-intensive process. The multiple round trips needed to fetch images occur very slowly in high-latency wireless networks. This process also quickly drains the battery, due to heavy CPU usage.

> If you do not care much about images, you can turn them off from Opera's Settings menu, which will speed things up considerably. Alternatively, Opera displays the page before it downloads all the images. So, you can press the Stop command to stop loading images if the text page is good enough.

The Opera Mobile Accelerator is a subscription-based service that can drastically improve Opera browser performance. The idea is to delegate much of the bandwidth- and CPU-intensive work from the mobile device to a *proxy server*. When you request a URL from the Opera browser, the request is forwarded to the proxy server, a server running on the Opera network that downloads web pages and images on your behalf. The proxy server fetches the web page and all the images in it, resizes the images, compresses all of them into a more compact format, and then returns the compressed package to the mobile browser. The mobile browser receives the optimized and compressed page and displays it.

It is easy to configure the accelerator proxy for your device. Just load the URL *http://www.opera.com/proxy* in Opera and the browser automatically configures itself. Behind the scenes, the browser is configured to talk to an HTTP proxy hosted by Opera (see Figure 9-8). Once the Opera browser is configured for the proxy, you get a 14-day trial period to try the service for free. Beyond that, you have to pay a monthly subscription to continue using the proxy service. The proxy server identifies your browser via the phone's International Mobile Equipment Identity (IMEI) number and verifies its subscription status. If you decide not to use the service at any time, just go to the Settings → Advanced menu and disable the proxy.

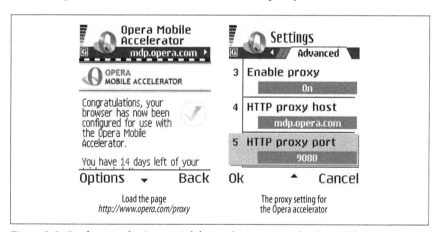

Figure 9-8. Configuring the Opera Mobile Accelerator proxy for the mobile browser

WebViewer: A Java-Powered Browser

The Opera browser is great, but it does not run on Nokia Series 40 devices. Even with accelerator proxy support, it still makes heavy use of the Series 60 device's network and CPU resources, since it still needs to parse and render the HTML code. If you want to browse HTML pages from a Nokia Series 40 device or use a browser that runs faster on a Nokia Series 60 device, you can choose Reqwireless's Java-based web client, called WebViewer.

The idea behind WebViewer is to push the browser proxy to the limit. The WebViewer client is only 48 KB, and you can install it on even first-generation Series 40 devices, such as the Nokia 7210. Without the proxy, the browser cannot render any web pages. In fact, the browser is configured to connect to the proxy only. The proxy not only fetches and resizes web content on behalf of the browser, but also parses the HTML and decides how to render the page. These rendering instructions, together with any embedded images and other media objects, are returned to the WebViewer browser in a highly compressed, proprietary binary format. This way, the device browser does not need to parse the HTML and manage the layout. It simply follows the well-formed instructions from the proxy to place text and images at specified places on the screen. The proxy renders the page to fit the device's screen size so that you will never need to scroll horizontally (similar to the SSR mode in the Opera browser). Figure 9-9 shows the O'Reilly home page rendered on the WebViewer browser on a Nokia Series 40 device.

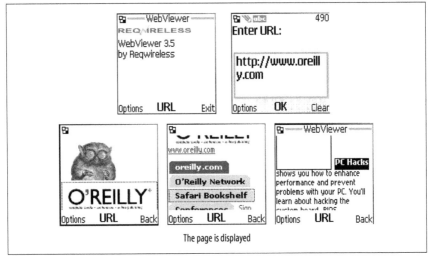

Figure 9-9. The O'Reilly home page displayed on a Series 40 device's WebViewer browser

By default, the WebViewer browser does not display images in web pages. You can enable images via the Settings menu.

In addition to HTML and common image formats (i.e., GIF and JPEG images), the proxy also can parse media formats that are not supported by most other browsers and can figure out how to render them on the devices. Those additional media formats include TIFF and BMP images, Microsoft Word and Excel documents, and PDF documents.

Of course, the proxy-based approach has its drawbacks. Unlike Opera, the WebViewer browser cannot function if you do not subscribe to the proxy service. If the proxy server is temporarily down, you will lose access to the Web. That said, I used both the WebViewer and Opera proxies regularly for a year or so, and never experienced any proxy downtime. Also, since Web-Viewer runs inside the Java sandbox, it is less integrated into the device-native software than the Services and Opera browsers. For example, although you can download and view full-size web images inside Web-Viewer, you *cannot* save them into the phone's memory.

The Mobile Web in Bite-Size Chunks

Mobile portals give you a launching-off point for your mobile web experience. But to make an intelligent choice of portals, you need to understand the core features and differences between portal services.

Accessing web sites directly via their URLs is OK if you have only a handful of web sites to visit. You can simply manage them in your browser's book-marks and load each of them by hand. However, if you need to keep track of many web sites, managing and loading them individually is cumbersome. Here is where web portals come in handy.

Portals and search engines are among the most frequently visited web sites on the Internet. A portal provides a single point of entry to access a wide variety of information on the Web. In the desktop PC world, the portal is primarily a content aggregator to help users track the ever-changing Internet. In the world of the mobile Web, however, portals play a much more important role than they do on the desktop:

- A mobile portal can aggregate content from regular, nonmobile Internet sources and present them in a format that is friendly to mobile browsers. For example, it can take a long story from the *New York Times'* HTML web site or a blogger's Really Simple Syndication (RSS) feed, break it into multiple smaller pages, and then feed it to the mobile browser as WML content.

- A mobile portal saves you from having to remember and type multiple URLs and usernames/passwords on a small phone keypad.

- With for-pay mobile content, a portal saves you the hassle of dealing with multiple content providers and billing services.

Most mobile portal solutions today are based on the concept of a "personalized portal." That is, first you register a personal account with the portal. Using a desktop PC, you can log into the portal's management interface and customize its contents. You specify what content your personal portal account will aggregate and display. Then, on the mobile phone, you see a personalized portal site that displays its contents according to your account settings. The separation between the management and view interfaces is a very smart design. You complete most of the interactions that require user input on the PC via the management interface. On your mobile device, you can stay in the comfortable "navigate and read" mode most of the time.

As you can imagine, the mobile portal is a great place to reach out to billions of mobile data users. So, many companies are in this space, competing for your business. In this hack, I'll introduce you to some of the most popular portals available via your Nokia phone.

The Wireless Operator's Portal

Your wireless operator already has a portal for you! If you got your phone from the operator (or an authorized reseller), the portal's URL is probably already defined in the Bookmarks section of your Services browser. In this section, I'll use T-Mobile USA's T-Zones service as an example. Other operators' mobile portals are similar.

For most operators, you need to register an account on the operator's web site to personalize the portal content. The T-Mobile USA web site is *http://www.t-mobile.com/*. Your account is associated with your phone number. You can check your bills, minutes, and service subscriptions from the web site. Of course, you also can customize your portal via the web site (see Figure 9-10). To get this customization page, click the T-Zones Settings link toward the top of the page after you log in.

Figure 9-11 demonstrates how to make changes to the movie listing settings when you click the Customize button. From the web-management interface, you can even preview the portal configured for the mobile device (see Figure 9-12).

Then, to access the T-Mobile portal from the phone's Services browser, enter the web address *http://wap.myvoicestream.com/*. On the phone browser, the portal page is divided into sections, and you can navigate through them via a series of menu-like links. Figure 9-13 shows how to get to the local movie listings from the phone browser. Notice that the theaters listed on the mobile portal page are the same ones configured in the management interface (refer back to Figure 9-10).

Figure 9-10. The portal content settings in a T-Mobile account

A key benefit of using the wireless operator's portal is that your device is automatically logged into the account, without requiring you to type a username and password on the keypad. How does that work? Well, recall that your portal account is associated with your phone number. Since the wireless operator knows where a General Packet Radio Service (GPRS) request originated (i.e., your phone number) at the WAP gateway, it can automatically log you into the portal.

Now, let's dig deeper into the content and services the T-Mobile portal offers. Other wireless operators have similar options.

Aggregated web content. The T-Mobile portal aggregates and displays the following web content on your phone browser:

- Movie listings and show times from your local theaters
- Your local weather forecast
- Quotes for stock and major fund prices

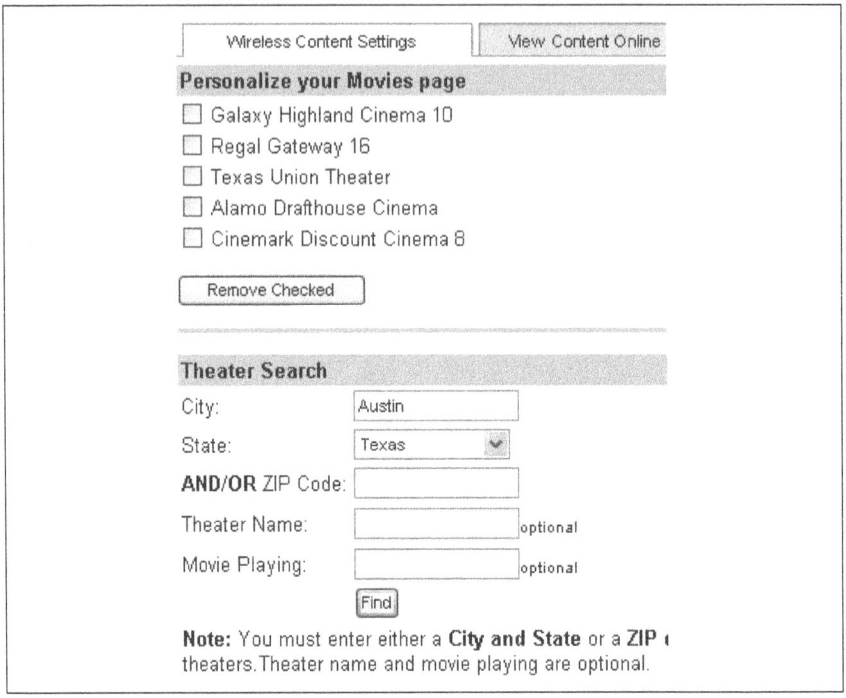

Figure 9-11. *Making changes to the movie listing content setting*

- Categorized news from various Internet sites
- Sports scores from your favorite teams
- Your daily horoscope
- Winning lottery numbers

The operator makes deals with the content providers so that you can view the information in the portal for free.

Bookmarks. The portal cannot possibly aggregate all content from all web sites. If you visit certain sites often, you can put their URLs in the Bookmarks section of the portal. That saves you from entering the URLs on the phone keypad from time to time. It is a lot easier to manage the long URLs on a desktop computer via the portal's web interface than it is to deal with them on the phone. Figure 9-14 shows how to manage the bookmarks on the T-Mobile web portal, and Figure 9-15 shows the bookmarks displayed on the phone browser.

Mobile content download. The portal provides downloadable content, such as wallpapers, screensavers, ring tones, themes, and Java games, specially tailored to your phone. You typically need to pay for each download. Most

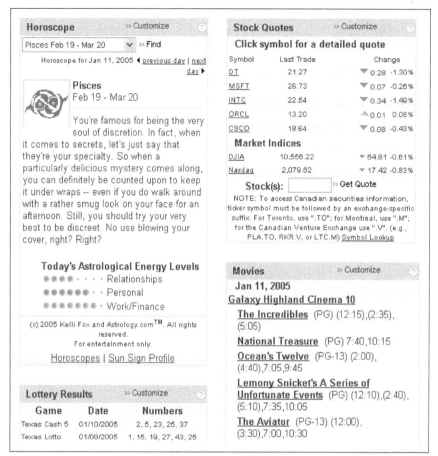

Figure 9-12. Previewing the customized portal contents from the web site on a desktop browser

operator portals support "one-click-shopping." In this case, the purchase is automatically billed to your monthly phone service bill. There is no need to enter credit card and payment information on the small phone keypad. This integrated billing is one of the big advantages of operator portals compared with other types of portals (discussed later in this hack). Figure 9-16 shows the process for buying a ring tone and billing it to your phone bill.

You can also purchase mobile content from the desktop PC browser and then download them to the PC. "Customize the Idle Screen" **[Hack #46]**, "Customize Ring Tones" **[Hack #45]**, and "Run Java Applications" **[Hack #16]** cover how to install wallpaper, ring tones, and Java applications, respectively, from a PC to a phone.

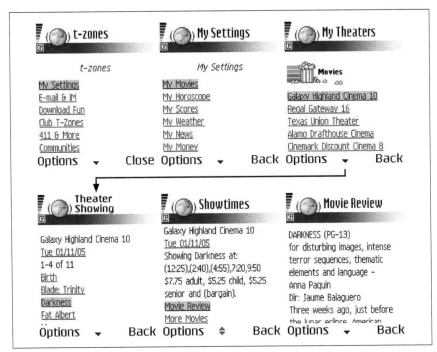

Figure 9-13. Finding a movie through the T-Mobile portal

The operator-assisted services. Operator-assisted services are very useful for mobile phone users. They let the operator search for information you're interested in and give the results to you via voice. For instance, if you are in a big city looking for a restaurant you know only by name, it would be impossible to search for it on your mobile phone while you are navigating the busy streets. It would be nice to have an operator look it up for you and give you driving directions over the phone.

 The operator-assisted services usually cost you between $1 and $2 per minute.

The operator-assisted services are also known as 411 services, since you typically dial 411 or similar service numbers on your mobile phone to reach the operator. As the range and the popularity of the services grow, other convenience numbers are also being provided for faster and more direct access to specific services. For instance, you might dial different numbers to have the operator search for a local person or a local business. The wireless operator's portal site often lists those service numbers and their per-minute charge rates. From the convenience of your Services browser, you can click one

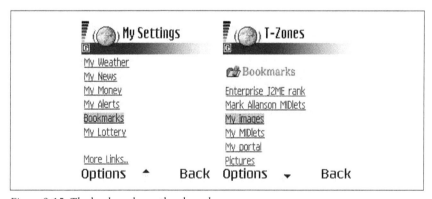

Figure 9-14. Adding a bookmark via the web account

Figure 9-15. The bookmarks on the phone browser

button to make the call and have it billed to your monthly bill (see
Figure 9-17). I discuss how to make calls from a web page in more detail in
"Make Phone Calls from Web Pages" [Hack #54].

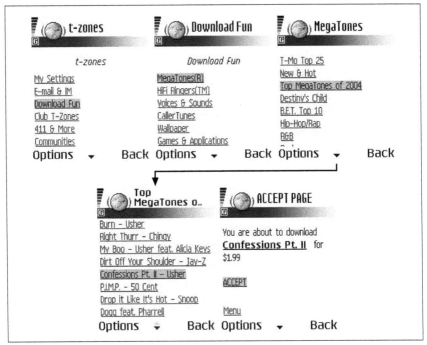

Figure 9-16. *Downloading and buying a ring tone from the portal*

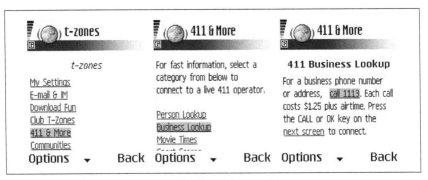

Figure 9-17. *Making operator-assisted calls via the portal*

Email, alerts, and instant messaging. In addition to delivering content accessible via mobile and desktop web browsers, portals also provide a variety of messaging services to the phone (I cover the details of those services in other hacks):

- You can access existing email accounts from your phone browser via the portal **[Hack #60]**.

- You can configure the portal to send alerts (e.g., birthday reminders and news headlines) via Short Message Service (SMS) messages **[Hack #64]**.

- You can interact with popular instant messaging systems via the portal [Hack #66].
- You can send new contact and calendar items from the portal to the phone via messages [Hack #34].

Most operators charge for each SMS message sent and received on your phone. If you use those messaging services frequently, the SMS messaging cost can add up. I recommend you check your operator's special rate plans for frequent SMS users.

Third-Party Portals

The wireless operator's portal is a free service to subscribers, and it integrates very well with your phone and service plan. You should take advantage of it. However, if you switch to another operator sometime in the future, you will probably lose all your personalized settings and customizations in the portal. Also, if you use a prepaid access card, you probably do not have access to the operator's portal at all.

To avoid operator lock-in, you can use one of the third-party mobile providers, such as Yahoo! Mobile, MSN Mobile, or AOL MyMobile. They offer the same types of services as the wireless operator's portals, with the addition of some premier content such as news analysis, driving directions, and Yellow Pages listings. Table 9-1 lists the URLs for the mobile sites and desktop management interfaces for those portals.

Table 9-1. Third-party mobile web portals

Name	Mobile URL	Portal management URL
Yahoo! Mobile	*mobile.yahoo.com*	*www.yahoo.com*
MSN Mobile	*mobile.msn.com*	*www.msn.com*
AOL MyMobile	*mymobile.aol.com*	*www.aol.com*

If you are already a Yahoo!, MSN, or AOL user, you can carry your existing desktop portal settings over to your phone. The third-party portals provide better integration with their respective email and instant messaging services. Since the third-party portals cannot securely identify your mobile phone browser by its phone number, you have to sign in from your phone to use them.

Custom Portals

All the portals I discussed so far allow you to choose content from their predefined pools of resources (a.k.a. their walled gardens). You cannot add more information sources, such as alternative news and real-time stock quotes, to those portals. The best you can do is to add a bookmark, which is inadequate in many cases. Some web sites allow you to build your own portals from a larger variety of content sources. Here are two examples:

WINKsite (http://www.winksite.com/)

Specializes in syndicated content from news and blog sites. You can set up a personal WINKsite portal and aggregate third-party content to your personal portal page. You can also use WINKsite to run surveys, chat rooms, and forums from your mobile portal page. Figure 9-18 shows how to configure your own WINKsite portal. I discuss how to use WINKsite to access blogs via your mobile phone in more detail in "Read Blogs and RSS Feeds" **[Hack #55]**.

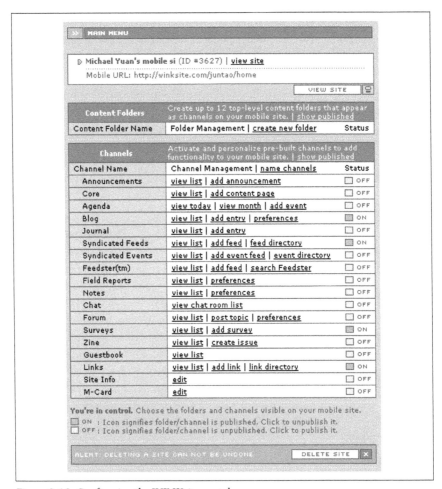

Figure 9-18. Configuring the WINKsite portal

Moreover (*http://www.moreover.com/*)

> Provides a large pool of major news organizations (e.g., the *New York Times* and CNN), online news sites (e.g., CNET and Salon), and regional news sources (e.g., major local newspapers and TV stations). It monitors business news, press releases, research reports, blogs, and even discussion boards. You can purchase and aggregate any of these on your portal.

Of course, you can always build your own mobile portal site from scratch if none of the existing ones satisfies your needs. Check out "Create a Mobile Web Site" **[Hack #53]** for more details.

Find What You're Looking For

Search engines allow you to quickly access information on the Web. They work just as well on the mobile Web if you know how to use them.

Search engines are the best tools for keeping track of the ever-changing landscape of the Web. You can think of a search engine as being a giant portal—one with the entire Web as its "content provider portal." The best search engine on the Web, Google, also has a mobile search engine specially designed for mobile phones. You can reach it via the URL *http://www.google.com/wml* from your Nokia phone's Services browser. The Google WML search engine allows you to search either the entire Web with all HTML web pages (see Figure 9-19), or the mobile Web with WML pages only.

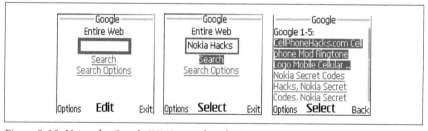

Figure 9-19. Using the Google WML search web site

The cool thing about Google's wireless search is that it supports a "number" input mode, whereby it guesses search query words from the numbers you enter. For example, to enter the word "book" in the normal input mode, you need to press 2 twice, press 6 three times, press 6 three times again, and finally press 5 twice. But with the Google number mode, you can simply type in the number 2665. Of all the possible combinations from those four digits, Google will figure out that "book" is a common word, and you probably intend to search for it. The Google number mode is very similar to the T9 input method [Hack #58], except that Google processes the translation on a backend server and has a much larger word dictionary.

 In fact, the Google wireless search page itself is accessible via the URL *http://www.466453.com/*. The numbers 4, 6, 6, 4, 5, 3 correspond to the letters g, o, o, g, l, e on a mobile phone keypad. It is T9 in the URL! This all-number URL is very fast to type on a keypad and is easy to remember.

Figure 9-20 shows how to access the Google number search web page, set the search option to Number Mobile Web, and then search the phrase "2665." The returned results are WML pages related to "book." (If Google insists on loading in full mode, you can force the small-screen version by going to *http://www.466453.com/wml*.)

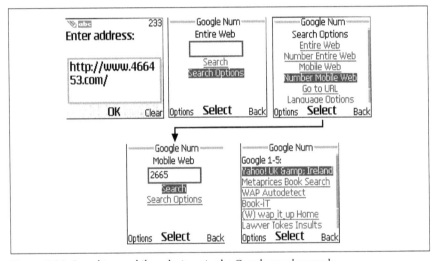

Figure 9-20. Searching mobile web sites via the Google number mode

Create a Mobile Web Site

Personal mobile web sites are a very effective means of sharing information and mobile media content. There are a few things you should know when authoring and deploying mobile web sites.

Many Internet hackers run their own web sites to communicate with peers. For advanced mobile hackers, setting up your own web site is a powerful way to enhance your mobile web experience. Via the web site, you can build a personal gateway to the Internet, share private content with family and friends, or even provide services to peers and the general public. In this hack, I'll first introduce the various mobile content markup languages with some simple examples, including an example of how to download and upload files from and to a web site for sharing. Then, I'll cover the server-side MIME-type configuration, which is necessary to make web pages accessible to most mobile phones, as well how to password-protect your data. Near the end, I'll provide a brief overview of mobile web development tools from Nokia.

The Web Made Small

To understand the various standards and specifications in the mobile Web landscape, you should learn a little about how mobile browsers and their content markup languages have evolved in the last several years. As I mentioned in "Browse the Web" [Hack #50], HTML, WML, and XHTML MP browsers are the ones you're most likely to find on Nokia devices.

HTML. Most Internet web pages for PC browsers are authored in HTML. However, in the early days of the mobile Internet (in the late 1990s), HTML was considered too heavyweight for most mobile phone browsers, for these reasons:

- The rich set of presentation elements supported in HTML is overkill for many mobile devices, especially low-end devices, since the phone screen simply cannot distinguish many font styles and media objects, and it provides limited room for layouts (e.g., frames and tables).

- Most HTML pages are not well-formed XML documents. For example, you can use the <p> tag to open a paragraph without the companion </p> to close the paragraph in HTML, but the software that draws your web pages on the screen has to check the rest of the HTML document to make sure there isn't a stray </p> it should pay attention to. This extra processing is nothing for the powerful CPUs in today's computers, but it can strain the limited CPU in your mobile phone. So, it is considerably more difficult to process irregular HTML than it is to process a well-formed XML document. Desktop web browsers have to perform all sorts of contortions to deal with permutations of HTML tags that have evolved since the early Web.

- After years of browser wars on the PC side, the HTML specification is fragmented with multiple proprietary extensions.

Overall, I do not recommend that you use HTML to create mobile pages, due to the limitations of current mobile browsers.

WML. Phone manufacturers and wireless operators developed their own lightweight content markup languages to replace HTML. Examples of such markup languages include cHTML, which is used by i-mode services, and HDML, which is promoted by Phone.com. However, having to deal with multiple markup languages was a big burden for mobile content developers and it hindered the adoption of the mobile Internet in those early days. To solve this problem, a standard mobile content markup language supported by all mobile device manufacturers and operators was needed. The Wireless Markup Language (WML), defined by the WAP Forum, emerged as such a standard.

> The WAP Forum is an industry-standard body that develops the data communications protocols for mobile networks. WML is officially part of the Application Environment specification of the WAP standard. Put simply, all WAP-compatible devices should support WML.

Unlike page-based HTML, a WML document is conceptualized into "a deck of cards." Each card represents one screen of content, and the internal links among the cards enable navigation from screen to screen. The ability to download multiple cards at once helps to reduce the slow and unreliable network round trips in WAP applications. The following code snippet demonstrates a simple WML document. The first card asks for your name and the second one echoes it back to you. The device needs to download the WML document only once to get both cards.

```
<?xml version='1.0'?>
<!DOCTYPE wml PUBLIC "-//WAPFORUM//DTD WML 1.2//EN"
         "http://www.wapforum.org/DTD/wml_1.2.xml">
<wml>
  <card id="Name" title="Enter Name">
    <do type="accept" label="SayHello">
      <go href="#Hello"/>
    </do>
    <p>Please enter your name:
      <input type="text" name="name"/>
    </p>
  </card>

  <card id="Hello" title="Say Hello">
    <p>Hello, $(name)</p>
  </card>
</wml>
```

> You've probably already given in to the temptation to put this document on a web site with the extension *.wml* and try it out with your phone. However, if your web server sends it with a Content-Type of *text/plain*, it might not load properly. To be sure your web server is sending it correctly, see "Configure the Server for MIME Types," later in this hack.

WML has made enormous progress in standardizing mobile browsers. However, as the mobile phone's capability improves, the need to keep the content markup language as light as possible has diminished. Instead, the challenge is to add more features to the WML standard and, at the same time, improve interoperability with the vast number of web sites in the wired Internet world.

XHTML MP. In WML 2.0, WML became a subset of the standard XHTML known as the XHTML Mobile Profile (XHTML MP). Since XHTML is a strict XML definition of HTML, it is a lot easier to process on resource-limited mobile devices (since it doesn't have to deal with all the exceptions allowed in the more permissive HTML used by many web sites). The following code snippet shows a simple XHTML MP document. Notice that it is also a well-formed XML document (it includes the closing tags that are not necessary in non-XHTML HTML documents).

```
<!DOCTYPE html
  PUBLIC "-//W3C//DTD XHTML 1.0 Strict//EN"
  "http://www.w3.org/TR/xhtml1/DTD/xhtml1-strict.dtd">
<html>
  <head>
    <title>An example XHTML page</title>
  </head>
  <body>
    <p>
      <b>Hello</b>
      <br/>
      <i>World</i>
    </p>
    <p>
      From your friends
    </p>
  </body>
</html>
```

XHTML is not yet the dominant markup language for Internet and mobile web sites. But it is widely supported in the new generation of authoring tools and dynamic web site frameworks. As those tools are being adopted and web site developers are more consistent about developing for multiple browsers, you can expect XHTML to be the convergence point of the desktop Internet and the mobile Web.

Embedded scripts. So far, I have covered static markup languages. In the real world, web pages often have a scripting component to add some interactivity on the client side. For example, in a user registration page, an embedded JavaScript can alert you to missing fields or incorrectly formatted fields before you submit the form. Most HTML browsers, including Opera, have the option to support JavaScript. For WML pages, WMLScript is the perfect tool to add interactivity.

Web-Based File Transfer

One of the primary reasons to set up a personal mobile web site is to share mobile content (e.g., wallpapers and ring tones) with friends, or across several devices you own. It is easy to set up a page where your phone can download files via its WML or HTML browsers. The native Services browser allows you to download common media format files to the Gallery. Most Nokia phones also support web-based file uploading. The following XHTML MP page shows a file upload form on the device browser. You can select a file from the Gallery and submit the form to the server (see Figure 9-21).

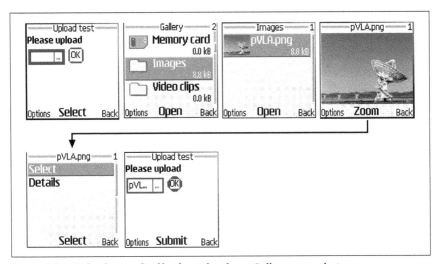

Figure 9-21. Uploading media files from the phone Gallery to a web site

The following listing is the HTML code for the file upload page:

```
<html>
  <head>
    <title>Upload test</title>
  </head>
  <body>
    <h3>Please upload</h3>
    <form action="processor.php" method="post">
      <input type="file"/>
      <input type="submit" name="submit" value="OK"/>
    </form>
  </body>
</html>
```

On the server, the following PHP script (i.e., the *processor.php* file on the web server) saves the file to the specified directory:

```
<html>
  <head>
    <title>Upload results</title>
  </head>
  <body>
<p><b>The file uploading result is:

<?php
  $uploadfile = '/upload/' . basename($_FILES['userfile']['name']);

  if (move_uploaded_file($_FILES['userfile']['tmp_name'],
                        $uploadfile)) {
    echo "<p>File was successfully uploaded.\n";
  } else {
    echo "<p>Invalid file!\n";
  }
?>
  </body>
</html>
```

> In PHP versions earlier than 4.1.0, you should use the $HTTP_POST_FILES variable rather than the $_FILES variable to access the uploaded file.

Configure the Server for MIME Types

After you write the web pages, it is time to publish them online. You can publish through either a web-hosting service or your own server. But before anyone can view your web pages, you must configure the server with the right Multipurpose Internet Mail Extensions (MIME) types.

When the web browser accesses any resource (e.g., a file) on the server, the server first returns the content's MIME type. Then, based on the MIME type, the browser determines how to render the content. For example, the server returns a *text/html* MIME type for an HTML page so that the browser knows it needs to parse the tags and render the HTML presentation elements. If the server returns the *text/plain* MIME type for the HTML page, the browser thinks it's plain text, and displays all the characters including the HTML tags. The server returns the *image/jpeg* MIME type of a JPEG image and the browser renders the binary data stream into an image. For the mobile browser to properly render the contents of your web site, the server must be configured to send the right MIME types. The server typically associates a MIME type for each filename suffix.

Scripts will generally set the correct MIME type using a
Content-Type header that's sent out in the HTTP response
header. So, the filename suffix of a script does not necessarily determine its MIME type. See "Dynamic content," later in
this section.

Static web pages. In the Apache server, the MIME type and filename extension associations are configured in the *conf/mime.types* file. Make sure the
following entries are available. Those entries specify the MIME types for
HTML, WML pages, WMLScripts, themes, Java and Symbian applications,
and video and audio files.

```
application/vnd.wap.wbxml           wbxml
text/html                           html htm
application/vnd.wap.wmlc            wmlc
application/vnd.wap.wmlscriptc      wmlsc
image/vnd.wap.wbmp                  wbmp
text/vnd.wap.wml                    wml
text/vnd.wap.wmlscript             wmls
video/3gpp                          3gp 3gpp
text/vnd.sun.j2me.app-descriptor    jad
application/java-archive            jar
application/vnd.symbian.install     sis
audio/midi                          mid midi kar
application/vnd.nok-s40theme        nth
```

The *application/vnd.wap.wmlc* MIME type refers to compiled WMLScript files. As I mentioned earlier, you need to
precompile the WMLScript if your phone does not use a
WAP gateway to access WML pages (most phones do). For
instance, you might access the WML page via a direct HTTP
link from your Internet access point rather than from the
WAP access point.

If your device browser fails to render or process any of the pages or embedded media elements from your server, you should first check the MIME type
settings for that page or element. If you share a web server with other users
and do not have access to the *conf/mime.types* file, you can create a *.htaccess*
file in your web directory and add the following entries:

```
AddType application/vnd.wap.wbxml         .wbxml
AddType text/html                         .html
AddType application/vnd.wap.wmlc          .wmlc
AddType application/vnd.wap.wmlscriptc    .wmlsc
AddType image/vnd.wap.wbmp                .wbmp
AddType text/vnd.wap.wml                  .wml
AddType text/vnd.wap.wmlscript            .wmls
AddType video/3gpp                        .3gp
```

```
AddType text/vnd.sun.j2me.app-descriptor    .jad
AddType application/java-archive            .jar
AddType application/vnd.symbian.install     .sis
AddType audio/midi                          .mid
AddType application/vnd.nok-s40theme        .nth
```

The *.htaccess* file lets you specify custom server configurations for each directory on your site.

 Some web-hosting providers disable support for the *.htaccess* file in user directories, for performance reasons. If that is the case, you need to contact your site administrator to add support for the appropriate MIME types.

If you use the Microsoft Internet Information Services (IIS), you can easily associate MIME types with files and directories using its graphical user interface (GUI) management tools. For detailed instructions, please refer to the following two online documents from Microsoft:

- For IIS 4.0 and 5.0, *http://www.microsoft.com/technet/prodtechnol/ windows2000serv/technologies/iis/maintain/featusability/mimeiis.mspx*

- For IIS 6.0, *http://www.microsoft.com/resources/documentation/iis/6/all/ proddocs/en-us/wsa_mimemapcfg.mspx*

If you use web servers other than Apache and IIS, please refer to your server manual to find out how to set MIME types.

Dynamic content. So far, I have discussed how to publish static mobile web content. But in many cases, a dynamic web site that performs some heavy lifting on the server is needed. By turning your mobile phone into the front-end for web applications, you can run sophisticated applications on the server side and have access to the core features anywhere, anytime. For example, you can use your mobile phone to manage a personal blog or database application that is too big to fit into your phone's memory. A lot of literature is available on the design and development of dynamic web applications, so I will not discuss that here. However, one thing that deserves special attention from mobile users is how to set up the correct MIME types for your dynamic pages.

For example, if you use PHP scripts, all the script files have the suffix *.php*. The server typically just sends out a *text/html* MIME type for the *.php* URL resources. You cannot use static mapping to associate the *.php* suffix with both HTML and WML MIME types. Hence, you must set the MIME type dynamically from inside the script. The following code snippets show how to set the MIME type from PHP, Perl, Java Server Pages (JSP), and ASP.NET scripts.

In PHP, you can use the header function to set the header values:

```
header("Content-Type: text/vnd.wap.wml; charset=utf-8");
```

In Perl CGI scripts, you can simply print the header information directly into the response stream before you write out the WML page data:

```
print "Content-Type: text/vnd.wap.wml; charset=utf-8\r\n";
```

If you are using the CGI module in Perl, do the following:

```
use CGI qw(:standard);
print header("text/vnd.wap.wml");
```

In JSP, you can use the page directive to set header values in the server response:

```
<%@page contentType="text/vnd.wap.wml;charset=utf-8"%>
```

In Microsoft Active Server Pages (ASP) or ASP.NET, you use the Response object's attributes to set the headers when the server generates the response data:

```
Response.ContentType = "text/vnd.wap.wml"
Response.Charset = "utf-8"
```

For advanced developers, many of the leading web application frameworks, such as Java Server Faces, Java Portals, and the ASP.NET Mobile Internet toolkit, support HTML, WML, XHTML MP, and many other mobile markup languages out of the box. What does that mean? Well, that means you need to develop your content using only an abstract set of presentation widgets (e.g., buttons, labels, and text boxes). The framework detects the browser at runtime based on information embedded in the HTTP request header. If it is an HTML browser from a PC, the framework transforms the widgets to HTML presentation tags. If it is a WML browser from a Nokia phone, the framework transforms the widgets to WML markup and sets the *text/vnd.wap.wml* MIME type. From the developer's point of view, you need to write the content only once, and it automatically becomes available on all devices.

Protect Your Content

The *.htaccess* file in the Apache server also allows you to use passwords to protect certain directories on the server. It comes in very handy for a personal web site to provision downloadable content for private use. The following lines in the *.htaccess* file specify that only users with the username myself can access the content in this directory and its subdirectories. The clear-text usernames and encrypted passwords are stored in the file */apache/password.file*.

```
AuthType Basic
AuthName "Personal Space"
AuthUserFile /apache/password.file
Require User myself
```

You can add username and password pairs to any password file using the htpasswd utility that comes with the Apache server. The following example adds the myself user to the /apache/password.file file:

```
# htpasswd /apache/password.file myself
New password: mypassword
Re-type new password: mypassword
Adding password for user myself
```

> If the /apache/password.file file does not exist yet, you can use htpasswd -c to create a new file.

The mobile browser asks the user to authenticate himself using his username and password to access a protected directory.

> In this example, I use HTTP Basic authentication to authenticate users. The username and password are transmitted in clear text over the Internet. For a more secure scheme, you need HTTP Digest authentication or an HTTPS connection.

You can place the password file anywhere on the server's local filesystem. However, you must not place it in directories that are accessible via public web browsers. A malicious hacker can fetch the password file and crack the passwords offline to gain access to your protected directories.

Nokia Browser Developer Tools

Forum Nokia provides several tools for developing and testing browser-based applications. All of these tools are freely available for download from the Tools section of the Forum Nokia web site (*http://www.forum.nokia.com*). The device emulators in the Series 40 and Series 60 Developer Platform SDKs emulate the phone's browser applications on a PC. You can start the device emulator from the SDK menu, navigate the browser application using the arrow keys, and enter the Internet address of your web page. Figure 9-22 shows the emulator in action. The SDK emulator is sufficient for most developers to test their web sites, without a real device. The SDK comes with a network traffic monitor so that developers can see exactly what is transported over the network to the emulator (see Figure 9-23).

The Nokia Mobile Internet Toolkit (NMIT), which is another free tool from Forum Nokia, provides an editor for WML and XHTML MP pages. You can

Figure 9-22. The Nokia phone SDK emulator

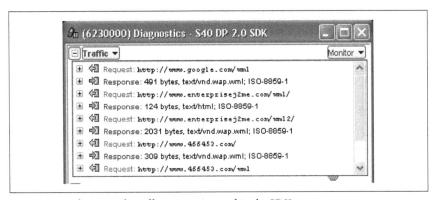

Figure 9-23. The network traffic monitoring tool in the SDK

use it to validate the tag syntax and then launch a browser emulator to test the page.

The Nokia WAP Gateway Simulator and Nokia WML Browser Emulator are two products that work together to emulate the entire end-to-end WAP infrastructure on the development PC. They are useful to developers who want to interact with the WAP gateway. Through the traffic monitor, you can see exactly how HTTP traffic is translated into WAP traffic, and vice versa.

Make Phone Calls from Web Pages

#54 You might have seen how you can click a link to dial your phone from your wireless operator's 411 service portal page. But how does this work? Well, the process is really easy, with a special WAP interface known as WTAI.

One of the coolest features in WML is its support for telephony functions known as the Wireless Telephony Applications Interface (WTAI). A Nokia WML browser allows users to make phone calls, send touch tones, and update the phone's Contacts list directly from a web page! Since the mobile phone is still primarily a voice communications tool, the ability to integrate the telephone experience with the web-browsing experience proves to be very useful. You can do that with either specially formatted URLs or WML-Script function calls.

Make a Phone Call

You can "link" to any telephone number via a specially formatted URL from your web page. Once you click that URL, the phone prompts you to make a call to the specified phone number. The URL can be embedded in a <go> or a <a> element in your WML, XHTML MP, or plain HTML pages. The following example shows an XHTML MP page with a phone call link:

```
<html>
  <head>
    <title>User Feedback</title>
  </head>

  <body>
    <center>
      <h3>User Feedback</h3>
    </center>

    <p>
      If you have any problem with our services,
      please feel free to
      <a href="wtai://wp/mc;+15555551234">give us a call!</a>
      Thank you very much.
    </p>
  </body>
</html>
```

The number +15555551234 after the wtai://wp/mc; string in the URL specifies the phone number. You can use any phone number format that your phone and wireless operator can understand.

Figure 9-24 shows that when you click the link and confirm the message, a call to the specified number is initiated. Once the call is connected, the phone goes back to the browser and you can continue browsing while staying on the phone. The small phone icon at the top-right corner of the last screen indicates that a voice call is in progress. Although the screenshots are taken from a Series 60 device, the process is similar on Series 40 devices. On a Series 40 device, you are given an additional choice to quit the browser before the call is initiated.

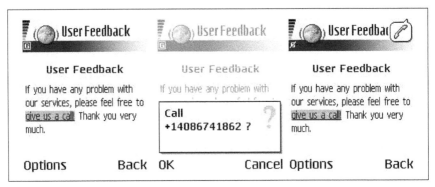

Figure 9-24. Making a call from the web page

 If you are browsing over a GPRS connection, the connection is suspended while the phone call is active.

Decide Whom to Call

In the previous example, the phone number is hardcoded into the URL. This is not very convenient in many scenarios. For instance, you might want a drop-down list of local numbers to choose from, or you might want to be able to enter your own target numbers. With a little help from WML variables, you can easily make the call link dynamic. The following example shows a WML text entry box. You can enter any phone number in the box and then click the link below it to make the call (see Figure 9-25).

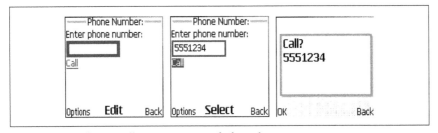

Figure 9-25. Making a call to any user-specified number

```
<wml>
  <card id="Call" title="Phone Number:">
```

```
    <p>
      Enter phone number:
      <input type="text" name="phoneno"/>
    </p>
    <p>
      <a href="wtai://wp/mc;$(phoneno)">Call</a>
    </p>
  </card>
</wml>
```

In the example, the phoneno WML variable captures your input in the text box. Then its value, $(phoneno), is embedded in the phone call URL as the phone number. If you enter an incorrectly formatted phone number (e.g., alphabetic letters with numbers), the browser will throw an error when you try to make the call.

Capture the Call Status

With the help of WML variables, you can also programmatically capture the status of calls made from any URL. The general form of the WTAI "make call" (i.e., mc) URL is as follows:

```
wtai://wp/mc;[phone number]![variable name]
```

The WML variable following the ! stores the call status code. Table 9-2 lists the possible status codes and their meanings.

Table 9-2. The status codes after the "make call" operation

Status code	Description
0	Success
-1	Unspecified error
-105	The other party is busy
-106	The network is not available
-107	The called party did not answer

In the following example, the user enters a number and makes the call. After the call is completed (or failed), the "Last call status" link at the bottom of the page opens a WML card showing the status of the last call.

```
<wml>
  <card id="Call" title="Phone Number:">

    <onevent type="onenterforward">
      <refresh>
        <setvar name="callstatus" value="No call has been made"/>
      </refresh>
    </onevent>

    <p>Enter phone number: <input type="text" name="phoneno"/></p>
```

```
<p><a href="wtai://wp/mc;$(phoneno)!callstatus">Call</a></p>
<br/><br/>
<center>
  <p><a href="#Status">Last call status</a></p>
</center>
</card>

<card id="Status" title="Status:">
  <p>The last call is to number: $(phoneno)</p>
  <p>Its status is: $(callstatus)</p>
</card>
</wml>
```

At the beginning of the card (page), I used a setvar element to initialize the WML variable. Figure 9-26 shows that the call cannot be completed, along with the corresponding status code.

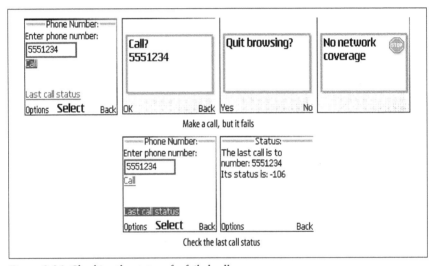

Figure 9-26. Checking the status of a failed call

Make Calls from WMLScript

The WTAI not only defines URL schemes for making phone calls, but it also defines public functions you can invoke in WMLScripts. WMLScript requires the WAP gateway (or precompilation), so it is not as easy to use as plain URL links. On the other hand, WMLScript allows you specific logical conditions for making calls, and it makes it easier to handle the call status return value. For more information on WMLScript programming, check out W3Schools' WMLScript tutorial at *http://www.w3schools.com/wmlscript/ default.asp*. The following WMLScript statement makes a phone call to number 5551234 and stores the return value in the flag variable:

```
var flag = WTAPublic.makeCall ("5551234");
```

If the call is successful, the return value in flag is an empty string. If the phone number is not formatted correctly, the flag value is invalid. If the call fails, the return value is one of the error codes in Table 9-2.

The makeCall function call blocks, and it will not return until the call is finished. Hence, your browser might freeze when you are making a call via WMLScript.

Other Telephony Functions

Besides initiating voice calls, WTAI public library URLs and WMLScript functions can also send touch tones and manipulate the phone Contacts list.

Send a DTMF tone. During an active voice call, if you click the following link, the phone will send touch tones (a.k.a. DTMF tones) for the keys 1#23*456 over the voice line:

```
<a href="wtai://wp/sd;1#23*456">Send tone</a>
```

This can be very useful for interacting with automated answering services, which ask you to press keys to navigate menus. You can also capture the return status of the operation in a WML variable. The following is the complete syntax of the "send DTMF" URL link. Table 9-3 lists the possible values of the status variable.

```
wtai://wp/sd;[tone sequence]![variable name]
```

Table 9-3. The status codes after the "send DTMF" operation

Status code	Description
0	Success
-1	Unspecified error
-108	There is no active voice connection

The corresponding WMLScript function for sending DTMF tones is as follows:

```
var flag = WTAPublic.sendDTMF ("1#23*456");
```

Add a phone book entry. On your web site's Contact Us page, you probably want to provide an easy way for visitors to add your contact number into their Contacts list without quitting the browser. The Add Phone Book Entry WTAI URL allows you to do just that. The following link prompts a user to

add the phone number +18001234567 as ABC Corp in their phone Contacts list (see Figure 9-27):

```
<a href="wtai://wp/ap;+18001234567;ABC Corp">
  Add our contact number
</a>
```

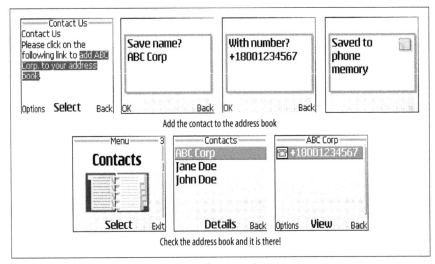

Figure 9-27. Adding a Contacts list entry from a web site

The complete syntax of the WTAI URL allows you to use a WML variable
to capture the return status value of the operation. Table 9-4 lists the possible return values.

```
wtai://wp/ap;[phone number];[contact name]![variable name]
```

Table 9-4. The status codes after the Add Phone Book Entry operation

Status code	Description
0	Success
-1	Unspecified error
-100	The contact name parameter is too long or unacceptable
-101	The phone number is invalid
-102	The phone number is too long
-103	The phone entry cannot be written
-104	The phone book is full

The corresponding function in WMLScript is addPBEntry:

```
var flag = WTAPublic.addPBEntry ("+18001234567", "ABC Corp");
```

So far, I have covered the WTAI public library URLs and functions, which all Nokia devices support. Some phones might support extended WTAI libraries, which provide functions for call management, SMS texting, and more Contacts list operations. You need to check your phone manual and the Forum Nokia web site (*http://forum.nokia.com*) to find out exactly what WTAI calls are supported on your device.

Read Blogs and RSS Feeds

Aggregate and read blog posts from your mobile phone.

Really Simple Syndication (RSS) is an XML-based content publishing format for syndicating news content from multiple web sites and information sources. Most news sites (e.g., *cnn.com* and *sourceforge.net*) and blog sites produce RSS feeds so that other sites and portals can aggregate their headlines. For more information about RSS, please see *Content Syndication with RSS* (O'Reilly, 2003).

Of course, RSS is not limited to aggregating web content across several web sites. Many blogging suites put the full text of blog posts, rather than just headlines and links, in the RSS feeds. In this case, you can read blog entries via RSS alone, without actually loading the HTML web site. In fact, many blog readers do just that. While many web browsers do not support RSS, some specialized tools, such as NetNewsWire for Mac OS X and SharpReader for Windows, allow you to keep track of many RSS feeds and read them from a computer.

But how do you read RSS from a Nokia mobile phone? The Nokia default browser (i.e., the Services application) and other third-party browsers [Hack #50] cannot parse and render RSS. In this hack, I'll discuss ways to read RSS on your phone.

Use a Web Portal

The easiest way to read RSS content is to set up a portal site that converts any RSS feed to regular WML content, which is accessible from any mobile phone browser. You can use the free service provided by WINKsite (*http://www.winksite.com/*) for this.

Figure 9-28 shows how to configure the RSS feeds in a personal WINKsite portal account. You can put any number of RSS feeds there.

Figure 9-29 shows what those RSS feeds look like on a Nokia Series 40 phone's Services browser. Notice that long blog posts are divided into several pages for easy reading on a mobile phone screen.

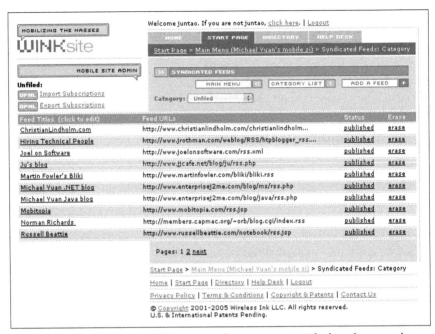

Figure 9-28. Configuring the WINKsite portal to aggregate RSS feeds and convert them to WML for mobile phones

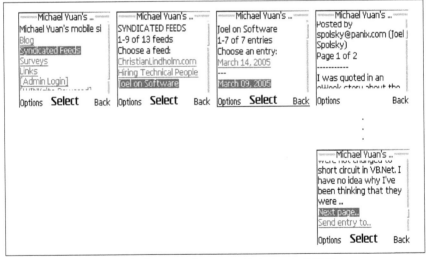

Figure 9-29. Reading RSS feeds on a Nokia Series 40 device via WINKsite

J2ME-Based RSS Reader

You can also download and install third-party programs to aggregate and render RSS content on your phone. The mReader program, written by Mark Allanson, is a J2ME-based RSS reader that works on Nokia Series 40 and Series 60 devices. It is available for free download from *http://markallanson. net/html/technical/j2me/mReader.htm*. mReader parses RSS content as well as basic HTML-formatting tags embedded in the RSS stream. Figure 9-30 shows the mReader program in action on a Series 60 smartphone.

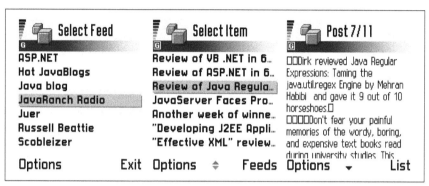

Figure 9-30. The mReader program in action

Typing URLs for RSS feeds on a mobile phone keypad can be tedious work. mReader supports importing a list of RSS feeds from an Outline Processor Markup Language (OMPL) file. You can create an OPML file by exporting RSS feeds from most RSS aggregator/reader programs. You can also download fully categorized RSS feed listings in the OPML format from directory web sites such as *http://www.w3os.nl/logos/opml/*.

Use a Web Proxy

The web portal and standalone Java client approaches have their benefits. It is easy to manage feeds on the web portal, and it is more economical, in terms of mobile phone bandwidth, to have the server polling the blogs for updates. The standalone Java client, on the other hand, provides a better user interface (UI) on the phone, which is crucial for successful mobile applications.

Litefeeds (*http://www.litefeeds.com/*) is a new service that combines the benefits of the web portal and the Java client approaches. Litefeeds provides a web site where you can create an account and edit your RSS feeds. The web site polls the RSS sources and keeps its content up-to-date at all times. On the mobile phone, Litefeeds distributes a Java client program that retrieves information from the web site as needed, using a compressed data protocol. The Java client presents the information in an effective mobile UI.

Post to Your Blog

#56 Use a mobile blogging service to post blog entries with pictures from your mobile phone.

Not long ago, bloggers could only update their blogs using a computer. Inspiration, however, does not always coincide with the presence of a bulky computer. At the turn of the century, some adventurous and creative bloggers started blogging from their mobile devices. The word *moblog* was thus coined, referring to blogging from a mobile device (mobile phones, PDAs, etc.).

If blogging without a computer is convenient, moblogging with a camera phone is exciting. In just a few clicks you can snap a quick shot and add a few punch lines, and minutes later the neatly formatted post on your personal blog can be shared with the entire world! Figure 9-31 shows a snapshot of O'Reilly editor Brian Jepson's moblog. You can see Brian's world through pictures taken from his phone camera and comments that captured his instantaneous thoughts. (That dry spell between November and March? That's when Brian used a phone without a camera for a few months before switching back to his Nokia 3650.)

There are two major approaches to moblogging, regardless of the myriad phone models and their different capabilities: SMS moblogging and email moblogging. This hack introduces and compares these two methods, with an emphasis on the latter.

> Instead of email, you can also use MMS to post blog entries from your camera phone. The MMS message is sent to an email address, and hence, MMS moblogging is essentially the same as email moblogging. See "Send Email the Easy Way" **[Hack #59]** for more details on the MMS-to-email gateways.

SMS Moblogging

SMS moblogging works on any handset that can send SMS, which is virtually every mobile phone nowadays. You write the blog entry on the mobile phone and send it via SMS to a service phone number provided by a moblog service. The moblog service interprets the received SMS message and posts it to your blog. You can be identified by the caller ID or an ID code embedded

Figure 9-31. An example photo moblog from Brian Jepson

in the SMS message. A good example of such an SMS moblog service is Txt-solutions (*http://blog.txtsolutions.com*). Txtsolutions can host your blog on its own site, or post your blog entries to your account in a list of supported third-party blog hosts.

> If the SMS moblog service posts the blog entries to a third-party blog host on your behalf, it needs to know your login information on that third-party site.

Despite the meager prerequisite of the handset, SMS moblogging has some major drawbacks. First is the nagging 160-character limitation for SMS. Second, even if you are laconic enough to squeeze each of your posts into fewer than 160 characters (not words), SMS moblogging is text-only. Third, you need to pay extra to the SMS moblog service provider. For example, Txtso-lutions charges £0.10 for each message. This charge is in addition to your regular SMS fee (from your wireless operator) and cost for the blog hosting service if it applies.

Email Moblogging

If your mobile phone supports sending email or MMS, email moblogging doesn't have any of the drawbacks of SMS moblogging:

- There is no practical length limitation, as long as your thumbs don't complain.
- You can post photos from a camera phone, adding richness to your post.
- There is no extra charge besides what you pay for the wireless data and regular blog hosting service.

 Please refer to "Send Email the Easy Way" **[Hack #59]** and "Send and Receive Email on Your Phone" **[Hack #60]** for tips on how email works on your Nokia mobile phone.

Email moblogging works in much the same way as its SMS counterpart, except that you send email from the phone to a (hopefully) secret email address. The email Inbox is polled by the moblog service at regular intervals (e.g., every five minutes). When a new message is detected, the moblog service retrieves the message and parses the content. If the moblog service supports picture attachments (most do), you can even have a mobile photo blog!

 It is important that you keep the email address private. Any message sent to that email address will be automatically posted on your blog.

The easiest way to set up an email-based moblog is to use a blog-hosting service that supports email posting. It typically takes only a few minutes. After you open a regular blog account at the hosting service, there is usually a field under the user preferences screen where you can set the private email address. All emails sent to this address will be posted to the blog you just opened.

Most popular blog-hosting providers now support email posting from mobile phones, either free of charge (e.g., Blogger) or with a premium paid account (e.g., LiveJournal). Figure 9-5 compares several providers that emphasize moblog support.

Table 9-5. Moblog support in popular blog hosting services

Name and URL	Cost (USD/yr)	Photo attachment	Notes
Blogger http://www.blogger.com	Free	N	Free email posting feature. No photo attachment support.
LiveJournal http://www.livejournal.com	$25	Y	Email posting option available for paid account only. Photo attachments are supported.
TypePad http://www.typepad.com	$60	Y	Moblogging feature available for all account types; 50MB storage and 1 GB per month bandwidth for basic subscription.
Radio UserLand http://radio.userland.com	$40	Y	Cost includes desktop blog software and one year of hosting. Email attachment posting requires additional software; 40MB storage.

Besides the major blog providers, some new providers specialize in moblog hosting. Among them are Textamerica (*http://www.textamerica.com*) and Fotolog (*http://www.fotolog.net*). Both of them offer free accounts for users to blog directly from their mobile phones.

> The free accounts often come with the condition that the service provider can display advertisements on your blog pages.

Photo Moblogging to Any Blog Service

The aforementioned photo moblogging solutions tie you into specific blog-hosting services. But in many cases, you do not want this tie-in:

- You could be running your own blog server and do not use any of the blog hosting services (see "Host Your Own Mobile Photo Blog" [Hack #57] for more on this subject).
- Your favorite blog host might not provide photo moblogging features.
- You might not want to pay the extra money to add photo moblogging support.

Is there a service that is free and independent of your blog server? As it turns out there is, and it's called Flickr.

Flickr (*http://www.flickr.com*) started out as a photo storage provider. The one feature that makes Flickr really interesting from a moblog point of view is that it allows you to upload photos (to Flickr) by email and create a blog entry at your designated blog site. The blog site can be any of the blog hosts mentioned in the previous section, or even your own web site running popular blog server software.

The post includes the email content as its main body and an inline thumbnail, which is linked to the picture stored on Flickr. The thumbnail posting feature in Flickr is really cool and sets it apart from other services, especially if you have a camera phone that produces megapixel pictures. Without resizing to thumbnails, megapixel pictures would clutter up your blog and mess up the web page layout (e.g., sidebars, etc.).

The free account at Flickr allows you to upload up to 20 MB of photos per month. A subscription fee of $24.95 per year will push the quota to 2 GB per month. Considering most (mobile phone) moblog attachments are smaller than 100 KB, the free Flickr account is more than sufficient for even the most aggressive mobloggers. Combining the Flickr email posting service with a free blog-hosting service (or your own blog server), you can set up a powerful and free moblog with photo posting in no time!

Figure 9-32 shows the information Flickr needs to post to your blog site on your behalf. Obviously, it needs to know your site address and account credentials.

Figure 9-32. Confirming URL and account credentials of your blog site

Figure 9-33 shows the private posting email addresses Flickr provides to you when you open an account. One address is for adding pictures to your Flickr account, and the other is for adding the pictures and then posting thumbnails to the designated blog.

Uploading by email

Your Account /

When you upload photos by email, use the **subject line** to give your photo a **title**, and the **body** of the email to give it a **description**.

Each photo you upload will also inherit your **default privacy settings**.

We can also send you an email to **add this address to your address book** if you like.

✓ **Now you can blog your photos to My Moblog automatically.**

Send your photos to Flickr using this address:

what38involve@photos.flickr.com **RESET**

...and do this when they're sent:

Tag *all* your photos uploaded by email with:

[_____] [?]

SAVE

Send to Flickr *and* your blog using this address:

what38involve2blog@photos.flickr.com

(At the moment, you're posting to **My Moblog**. You can change to another blog, or blogging via email here if you wish.)

Figure 9-33. The free email address Flickr provides for each account to post blog entries and pictures

Smart client for Flickr. The default email application on some Nokia phones restricts the size of an attachment to a maximum of 100 KB. To get around this limitation and to provide a more integrated user experience, you can use a phone application that takes pictures and then automatically posts them to Flickr without having to switch applications and enter email addresses.

"Manoj TK" wrote a little open source J2ME program that does just that. You can download it from *https://sourceforge.net/projects/moblogger2/*. To learn how the program works, please refer to *http://manojtk.blogspot.com/2005/01/smtp-email-attachments-from-nokia-6600.html*.

—Haihao Wu

Host Your Own Mobile Photo Blog

Set up and run your own servers for moblogging!

Many power bloggers run their own blogging server so that they can have complete control over the content and presentation of their blog. While Flickr **[Hack #56]** can interoperate with many popular blog systems, it is still a "hosted" service that is beyond the blogger's control. For instance, there is no guarantee that the free Flickr service won't be replaced by a for-pay service in the future. If you want complete end-to-end control over your moblog, you can customize existing blog server software to make it support email posting with photo attachments from a mobile phone. In this hack, you will learn how to set up the popular WordPress blog server to support photo moblog postings.

> Running your own server also gives you complete access to web site visitor statistics, the ability to back up the web site, and many other features.

Set Up WordPress

WordPress (*http://www.wordpress.org/*) is an open source blog server based on PHP and MySQL. It is available free of charge from the WordPress web site. You can run it on your own server or on any Internet Service Provider (ISP) server account that supports PHP 4.1 and MySQL 3.23.23 or greater.

> As a PHP and MySQL application, WordPress runs on almost any modern operating system, including Linux, Unix, BSD, Windows, and Mac OS X.

By following the installation instructions (see *http://wordpress.org/docs/installation/5-minute/*) you can install WordPress in five minutes. After creating a user account, you can start to post and publish blogs on your very own blog server now! Out of the box, you can use Flickr to post photos to your freshly installed WordPress server **[Hack #56]**. In the rest of this hack, I will discuss how to set up WordPress so that you can moblog with it without needing a third-party service.

Email Posting with WordPress

WordPress supports email posting, albeit with two main drawbacks. First, it does not support email attachments. And second, it does not filter the sender's email address. Anyone can send email to the email address and have their messages appear on your blog. Without sender-address filtering, your moblog can be easily flooded with spam messages.

A patch by "lansmash" fixed both drawbacks. Applying the patch is straightforward. First, download the file *wp-mail-0.2.zip* from the web address *http://blade.lansmash.com/index.php?cat=5*. Unzip the archive to the WordPress root directory. This adds the files *PEAR.php* and *mimedecode. php* to the directory and replaces the original *wp-mail.php* file. Now, create two directories, *wp-photos* and *wp-filez*, under the WordPress root directory to store images and other attachments, respectively. You can choose other directory names as long as you remember to modify the corresponding settings in the new *wp-mail.php* file.

If you want automatic picture resizing (an important feature for megapixel camera phones), you need to install an additional hack provided by "Hugo." Simply replace the default *wp-mail.php* file in your WordPress installation with a new file downloaded from *http://www.vienna360.net/files/wp-mail.phps*.

Now, you can test your setup by sending an email message with a picture attachment to the email address specified in the Options → Writing section of the WordPress configuration page. Make sure you enter the correct server and login information. Nothing appears on your blog just yet. That is because the *wp-mail.php* script needs to be manually loaded to poll the email server for the new message. Do this by pointing your browser to *http:// your-domain-name/MyMoblog/wp-mail.php*.

Of course, if you need to open a browser and load the *wp-mail.php* script manually every time you post a moblog entry, it defeats the whole point of moblogging. So, let's discuss how to automate the email polling process in WordPress.

Set Up Email Automatic Polling

To automate the email polling and blog posting process in WordPress, you need to schedule a recurring task that loads the *wp-mail.php* script at fixed intervals (e.g., every five minutes). You need two tools for this: a lightweight URL loader called *curl* to load the PHP script, and the open source *cron* daemon to schedule the recurring URL loading tasks. Both *cron* and *curl* are installed by default on Linux, Unix, and Mac OS X computers. For Windows, you can install the Cygwin toolkit (*http://www.cygwin.com/*) to get a common set of Unix utilities including *cron* and *curl*.

 You can use a fully featured web browser, such as Firefox, to load the *wp-mail.php* script every five minutes. But that is a huge waste of server resources.

Next, log into your server and type the following on the command line:

```
crontab -e
```

This opens the *cron* control file, which contains all scheduled tasks, in the system's default text editor. You should append one line to the bottom of the file, save it, and exit the editor. *cron* automatically reloads the control file.

```
*/5 * * * * /usr/local/bin/curl http://yourdomain/wp-mail.php > /dev/null
```

The preceding line in the *cron* control file schedules the system to run the curl command in the last part of the line every five minutes in every year, month, week, and day. That's it! You now have a fully functional photo moblog server.

> If you do not want to use the default editor, you can always export the *cron* control file to a text file, edit it externally, and then import the edited file back to *crontab*. See the *crontab* manual page for more information.

—Haihao Wu

Email and Messaging
Hacks 58–66

Email is a killer application for the Internet. For business travelers, the ability to send and receive email on a mobile phone is a mandatory feature that certainly justifies expensive devices and data subscription plans. In fact, vendors such as Research In Motion (RIM) have built very successful platforms around mobile email with dedicated email devices and special subscription services. However, those dedicated devices and services are often too expensive for casual travelers. In addition, those dedicated email devices are often not as feature-rich as Nokia Series 40 and Series 60 phones. In this chapter, you will learn techniques for setting up and managing email accounts on your Nokia phone. Since one of the major obstacles of using email on Nokia phones is the lack of a keyboard, I will also include a hack that specifically covers how to type text messages quickly and accurately on the phone.

Compared with email, mobile messaging—including Short Message Service (SMS) and MMS—is much more ubiquitous on a phone. In this chapter, you will also learn innovative ways to use SMS and MMS to make the most out of your phone and services. To unify mobile messaging and computer messaging, techniques to access Internet instant messaging (IM) services from the phone are covered in the last hack of this chapter.

HACK #58 Type on the Small Keypad

Composing your messages on a phone keypad doesn't have to be painful. Most handsets make it pretty easy, but the easy way is sometimes obscure.

Typing on the small keypad is always the challenge for mobile phone messaging. Some Nokia smartphones, such as the 6800/6820 in Series 40 and 9300/9500 in Series 80, are equipped with fold-out QWERTY keyboards. Those devices are ideal for messaging and enterprise applications.

For Nokia Series 60 smartphones, it is often possible to use an external keyboard. The Nokia Wireless Keyboard is a full-size, foldable keyboard that works over Bluetooth. Only Nokia 7610 and newer Series 60 smartphones are officially supported by this keyboard, but Nokia 6600 is also known to work. Besides Nokia's own offerings, the Think Outside Stowaway Bluetooth keyboard (*http://www.thinkoutside.com/*) is another choice. The Think Outside keyboard supports older Nokia Series 60 devices such as the Nokia 3600 and 3650.

While keyboards are nice, most mobile phone users still need to rely on the keypad to enter messages. In this hack, I will cover several important techniques to improve text input on regular smartphone keypads.

T9 Text Input

When entering text on the phone's numeric keypad, most people rely on the easiest but slowest multitapping method. Using multitapping, each alphabetic letter requires multiple repeated keystrokes on the keypad. For instance, to enter the letter i, you need to tap the 4 key three times; to enter the letter c, you need to tap the 2 key three times; and so on. Hence, typing a message such as "I cannot find my pants" requires that you press the numbers 444 222266666668 333444663 6999 726687777 on your phone's keypad. Come on, people, the Morse code is faster!

> In the multitapping mode (alphabetic input mode), it is slow to enter numbers. For instance, you need to press the 2 key four times to enter the number 2. You can enter numbers quickly by switching to the numeric input mode, but switching modes is slow when you have mixed letters and numbers. To speed up number input, you can press and hold the number key to get that number without leaving the current input mode.

Thankfully your Nokia phone supports Text on Nine Keys (T9), which lets you type that phrase in roughly half the number of keystrokes: 4 226668 3463 69 72687. Figure 10-1 shows the sequence for typing "pants."

T9 is a predictive-text input engine from Tegic Communications (which is owned by AOL) for mobile devices such as your mobile phone. Instead of having to press each key multiple times for just one letter, you press each numeric key only once per letter; as you're pressing the keys, T9 will examine what you're typing and predictively figure out the potential combinations for valid words. Most handsets will pop up a list that shows you which word Tegic thinks you're typing, but you can just wait until you are done typing the word, and it will be right in most cases. If not, you can select the next matching word from the T9 dictionary. If you don't see anything marked "next," it is probably the 0 key.

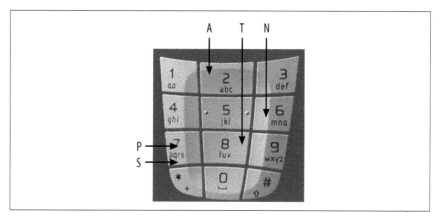

Figure 10-1. Typing with T9

 You might find that T9 is not enabled on your phone by default. On a Series 60 device, tap the pen/ABC key and a menu appears: select Predictive Text → On from this menu. (You can switch it off by selecting Predictive Text → Off from the menu.) On a Series 40 device, press and hold the Options soft key to toggle predictive input. You will know when predictive text is enabled by looking at the pencil icon at the top of your screen. A pencil with lines under it indicates predictive text, and no lines indicates multitap, as shown in Figure 10-2.

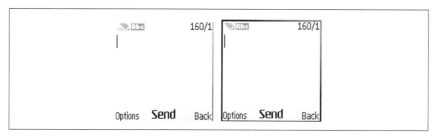

Figure 10-2. A Series 40 device with predictive text turned on and off

Most handsets also allow you to add words to your dictionary. When confronted with a word the dictionary doesn't know, the phone asks you to spell it, and then it saves it to its local dictionary. In the episode of *Seinfeld* called "The Soup Nazi," Jerry and his girlfriend call each other "shmoopie" as a term of endearment. Figure 10-3 shows me trying to add this word to the T9 dictionary on my Nokia 3650. The question mark at the end (and the appearance of the Spell soft key) indicates that T9 wasn't able to find the word in the dictionary. The "sion" in the figure is T9's guess as to what word I meant to look up.

Figure 10-3. Tegic encountering a new word

Since Tegic doesn't recognize the word, I need to type it in. First, I must press the */+ key to cycle through all the possible spellings. When I've gone through them all, the Spell soft key appears (it might temporarily take the place of the Options soft key). Next, I must press the Spell soft key to use multitap to input my word the old-fashioned way, as shown in Figure 10-4.

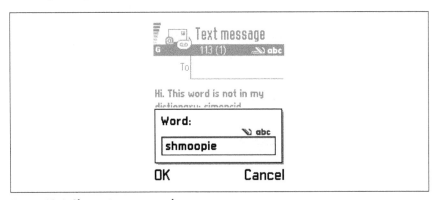

Figure 10-4. Shmoopie: not a word

I correctly spell the word "shmoopie" and press OK to save it for later. Now I can use T9 to type "shmoopie" whenever I desire (see Figure 10-5). Believe it or not, I use that word at least twice a day on workdays, and once per weekend.

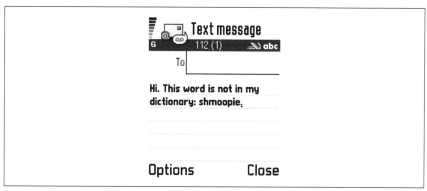

Figure 10-5. Shmoopie: now a word

Let's look at how long it would have taken me to type this with multitap. Before having this word in my dictionary, typing it out took 19 keystrokes, not including pauses between the "m" and "o" keys (when you use multitap to type two consecutive letters that appear on the same number key, you have to pause between typing letters). Now it takes only as many button pushes as the word contains letters—8, in fact, which takes a lot less time than 19, even if you're a really fast multitapper.

See *http://www.t9.com/* for more information and interactive demos.

Copy and Paste

One of the most overlooked items in the manual for your handset is the ability to copy and paste text. I found myself wishing I could do this on several occasions, blissfully unaware that this functionality is actually built into my phone.

On a Series 60 phone (Nokia 3650, N-Gage, and many others), simply press and hold the pen/ABC key when you are editing text in an email, the Note Viewer, or other application. Normally this button allows you to change input modes from various languages, or to switch between T9 and multitap. If you press and hold this button, however, any movement with the navigation pad (i.e., the joystick) will draw a selection box around the text of your choosing.

Once you have selected some text in this way, press the Options soft key and select the Copy option. To paste, simply put the cursor where you want it, press the Options soft key again, and select the Paste option, as shown in Figure 10-6.

Figure 10-6. Copying and pasting on a Series 60 device

This is a clever way to add address information from an email to the contact card of a friend, or to capture a long URL from your browser to email to someone else.

—Emory Lundberg

Send Email the Easy Way

Using the existing mobile messaging infrastructure to send and receive email messages is convenient and fits seamlessly with the mobile user experience.

The easiest way to send and receive email messages is to use the underlying wireless messaging infrastructure. You can use the SMS and MMS clients built into your phone to transport email messages. The tight integration between the message client and the wireless network offers some great benefits from the user's perspective:

- It is very easy to use. You do not need to install any additional software or configure anything on the phone or in the email account. It just works.

- The incoming messages are pushed to your phone. You are notified with a tone or an on-screen message when the message arrives. There is no need to push the "Check mail" button to check and retrieve messages.

Of course, this approach also has several drawbacks that you should be aware of when deciding whether this is the right way for you to connect to email:

- The biggest drawback is that you cannot use your existing email accounts to send or receive messages. Instead, the messages are routed through special accounts provided by the wireless operator. Please refer to "Send and Receive Email on Your Phone" **[Hack #60]** if you want to use your existing email accounts.

- Although the email accounts and infrastructure are free, you need to subscribe to MMS and/or SMS services, and there might be a per-message charge.

- There are size limits for both SMS and MMS messages. You cannot send long messages or big attachments. I will illustrate this point in more detail later in this hack.

- The wireless network does not have a service guarantee for SMS or MMS. It might take a long time to deliver a message. The message might be dropped silently if it is not delivered in 24 hours.

Overall, this method works great for casual mobile email users. Now let's check out exactly how it works.

> The regular email infrastructure over the Internet also does not have a service-level guarantee. But the wireless data network is less reliable than the Internet. In addition, the SMS traffic is low priority on wireless networks. Also, when an Internet email message doesn't make it to its destination, you will usually receive a failure notification.

Send Email Via MMS

The MMS service allows you to send messages with multimedia attachments from one phone to another, or from a phone to any email address. All you need to do is to open the Messaging application (Series 60) or the Messages menu (Series 40), choose New Message, and then choose to create a multimedia message. Using the Options → Add Recipient menu option, you can select any email address from your Contacts list in the To or Cc fields. In the message composition window, you can type some text and/or attach media files. Once you click Options → Send, the phone queues the message for delivery to the selected email addresses. Figure 10-7 demonstrates this process.

Alternatively, you can choose any media file in the Gallery and use the Options → Send → Via MMS menu item to create a message composition form with the file already attached. Or, you can choose any person from the Contacts list and use the Options → Create message → Multimedia message menu to create a message with the To field already filled in.

> Most wireless operators limit the size of MMS messages to 100 KB or less. You probably will be unable to attach more than two VGA (640×480) quality pictures in the message.

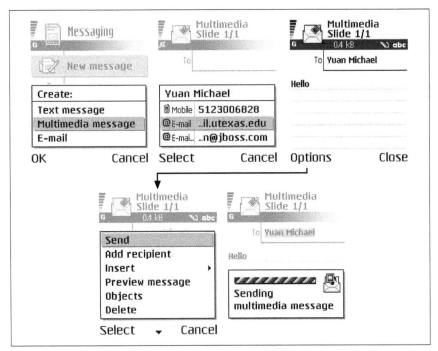

Figure 10-7. Composing and sending an MMS message to an email address

On your email client, you can retrieve the message and view its contents, including any attachments. Figure 10-8 shows the message from the phone. Your wireless operator might add a logo or some custom graphics to the message.

Receive Email Via SMS

If you look more closely at Figure 10-8, you'll see that the message comes from an email address that contains a mobile phone number. If you reply to this message or simply send a new message to that address, you will receive it on the phone as an SMS message (see Figures 10-9 and 10-10). This feature is known as an *email-to-SMS gateway*.

> Almost all wireless operators offer this kind of gateway at no extra charge beyond regular SMS subscription or per-message charges. An alternative way to send SMS messages via the wireless operator is to submit the message and target phone number to a web-to-SMS web form typically available on the operator's consumer portal web site once you log in.

Figure 10-8. *The MMS message received on a computer email client*

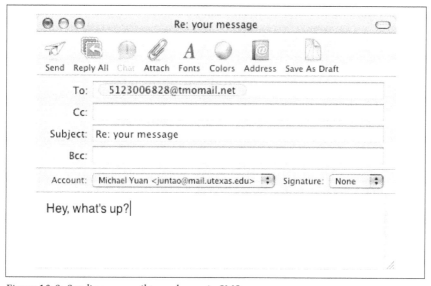

Figure 10-9. *Sending an email to a phone via SMS*

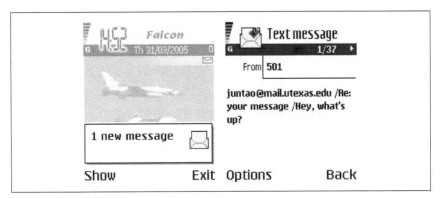

Figure 10-10. Receiving the email as an SMS message on the phone

Table 10-1 shows the web form and gateway email addresses for popular wireless operators in the U.S. If your operator is not in this table, you can typically call the operator's technical support to find out its gateway email address. This way, you can get your email messages delivered to almost any phone as SMS messages.

Table 10-1. Wireless operators' interfaces to SMS

Operator	Web	Email
Nextel	http://messaging.nextel.com/	<number>@messaging.nextel.com
SprintPCS	http://messaging.sprintpcs.com/	<number>@messaging.sprintpcs.com
AT&T Wireless	http://www.mymmode.com/messagecenter/	<number>@mobile.att.net
Cingular	http://www.cingular.com/sendamessage	<number>@mobile.mycingular.com
Verizon	http://www.vtext.com/	<number>@vtext.com
T-Mobile	http://www.t-mobile.com/messaging/	<number>@tmomail.net

In addition to the operator gateways, other commercial email-to-SMS gateways interoperate with multiple operators. For instance, a service called Teleflip (*http://www.teleflip.com/*) forwards any message sent to the address *<number>@teleflip.com* to the target phone number as an SMS message (this service works only in the U.S.).

The SMS message is limited to 160 characters. You cannot send long messages or message attachments to a phone via SMS.

Since SMS messages cost money to send and receive, it is crucial that wireless operators incorporate some security measures into the SMS-to-email gateway to prevent spam. As a result, you cannot send more than 10 messages in a short period of time (typically several minutes) via the gateway. To send many SMS messages from the PC or a backend server to a phone, please see "SMS from a Computer" **[Hack #63]**.

Send and Receive Email on Your Phone

Send and receive email on your phone using your regular email account.

To send and receive email via SMS and MMS **[Hack #59]**, you need to use the wireless operator's special email address, which is tied to your phone number. While this is convenient, it does not address the needs of most email users, especially business users, who need to manage email in multiple existing accounts while on the road. To access regular email on a Nokia phone, you have to configure the phone with your email server and account information. In this hack, I will teach you how to do that. But first, let's go over some basic concepts regarding the Internet email infrastructure.

To access email on your phone, you also need to subscribe to General Packet Radio Service (GPRS) or Code Division Multiple Access (CDMA) 1x data services. On some networks, the MMS access point provides access to email as well.

Email Basics

When you send and receive email messages over the Internet, your email client software on the PC or on the phone primarily needs to communicate with two types of email servers over the TCP/IP protocol:

Simple Mail Transport Protocol (SMTP) server
> Takes email messages from your email client and sends them to the recipient's email server. To avoid being exploited by spammers, most SMTP servers require you to have an account with them and to authenticate yourself before you can send an email message.

POP or IMAP servers
> Give you access to your email accounts. Your email client software logs into the POP or IMAP server periodically, using your username and password, to check for and retrieve new messages.

Typically, your SMTP and POP/IMAP servers are hosted by your Internet Service Provider (ISP) or by your company's IT department. If you use web mail (e.g., Yahoo! Mail and Hotmail), you typically need to pay for the privilege to directly access their mail servers.

 An open source program called YPOPs emulates a POP3 account over your free Yahoo! Mail account on your own computer. For more information, visit *http://yahoopops. sourceforge.net/*.

Use the Native Email Client

Nokia phones have a built-in email client in the Messaging application (Series 60) or the Messages main menu (Series 40). In Nokia terms, a pair of SMTP and POP/IMAP settings is a *mailbox*, since it provides a complete configuration for sending and receiving messages via an email address.

Set up your mailboxes. Setting up mailboxes is quite easy on both Series 60 and Series 40 phones.

On a Series 60 device, you can just launch the Messaging application and choose the Options → Settings → Email → Mailboxes menu. Then you can select and edit any existing email message in the Inbox. Or, you can use the Options → New mailbox menu item to create a new mailbox for an email account. The Messaging application's main screen lists the phone's native Inbox (for receiving SMS and MMS messages) as well as all configured mailboxes. Email messages are pulled into the mailboxes, not into the native Inbox. The Options → Settings → Email → Mailbox in use menu item points to the default mailbox to use when sending email messages.

A Series 40 phone can hold 5–20 mailboxes (i.e., email accounts). You can activate a mailbox via the menu path Messages → Message settings → Email messages → Active email settings (or via Messages → Email → Setting → Mailbox in use for older phones). The email Inbox (in Messages → Email → Inbox) on the phone retrieves messages from the currently active mailbox, and all outgoing email messages are sent via the SMTP server in the active mailbox. To edit the settings for a mailbox, you need to first activate it and then select the "Edit active email settings" (or "Edit active mailbox") menu in the email settings screen.

In the mailbox settings screen for both Series 60 and Series 40 phones, you need to enter the following information:

- Your email address
- An SMTP server address and its access credentials
- A POP/IMAP server and its access credentials
- The data access point to connect to the email servers

Figure 10-11 shows the settings for one of my mailboxes on a Nokia 6600 phone. The security setting specifies whether the phone uses encrypted connections (SSL) to connect the email servers. You should always try this option first, and disable it if you get an error. SSL protects your username and password from eavesdroppers. Although it also transports your email messages securely, the moment a plain-text email message hits the wide-open Internet, it becomes susceptible to eavesdroppers.

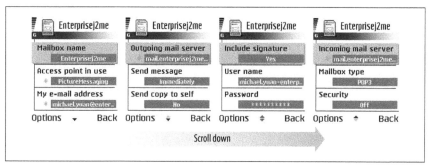

Figure 10-11. Email settings for a mailbox

In Figure 10-11, the SMTP server and POP/IMAP server share the same GPRS access point and the same username/password. Some devices (e.g., the Nokia 6230 phone in Series 40) allow you to specify a different GPRS access point, and a different username/password for the SMTP server and the POP/IMAP server, respectively. Many phones also support the Authenticated POP (APOP) protocol to retrieve messages. Under APOP, the username and password are sent in encrypted format to the POP server. APOP provides protection against network sniffers that intercept clear-text passwords.

Send a message. To create a new message on the phone, choose the Options → Create message → Email menu in the Messaging application (Series 60) or the Messages → Email → Create email menu (Series 40). You can select any recipient from the phone's Contacts list or type in any valid email address in the To and Cc fields. The body field of the email composition form is prepopulated with your email signature, if you have one. Type in your message and select the Options → Send menu to send the message using the currently active mailbox settings (see Figure 10-12).

> On Nokia Series 60 phones, you can also add nontext attachments to an email message. See "Handle attachments," later in this hack.

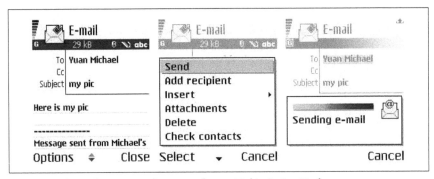

Figure 10-12. Sending an email message from a Nokia Series 60 phone

Receive messages. On a Nokia Series 60 device, you can simply open any properly configured email mailbox from the Messaging application's Main menu. The phone connects to the Internet and checks for new messages on the remote POP/IMAP server. Or, you can select Options → Connect to force a connection.

On a Nokia Series 40 device, select the Messages → Email → Retrieve menu to retrieve messages from the currently active mailbox. The retrieved messages are available under the Messages → Email → Inbox menu.

For all new messages, the phone retrieves only those message headers that show the sender's email addresses, the message dates, and the message subjects. It does not automatically retrieve the message bodies or attachments. This saves bandwidth and greatly improves the email client's response time. If you want to view an individual message, you can retrieve the message body by opening it (see Figure 10-13). You can also select commands from the Options menu to retrieve all messages or marked messages.

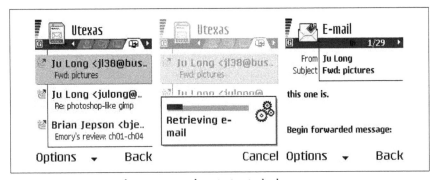

Figure 10-13. Opening the message and retrieving its body content

Delete messages. Since the Nokia phone first retrieves all the message headers from the remote POP/IMAP server, it retrieves the message bodies only when you specifically instruct it to do so (e.g., by opening a message). So, when you delete a message via the Options → Delete menu, you have two options:

- If you choose to delete from the phone memory only, the message body is deleted. But the message header remains in the phone memory. You can still see the message in the email message list, and if you open it, the phone redownloads the message's body.

- If you choose to delete the message from both the phone and the server, the message header is marked with a delete sign in the phone's email list. The next time the phone connects to the server, it instructs the server to delete the message immediately.

Figure 10-14 shows the process to delete a message from both the phone and the server on a Nokia Series 60 phone.

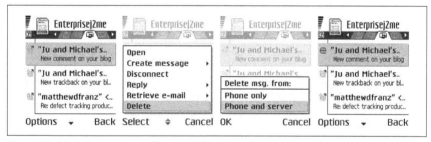

Figure 10-14. Deleting an email message from a Nokia Series 60 phone

When deleting a message, the Nokia email client behaves differently from most desktop email clients. Desktop email clients can be configured to delete the message on the server several days after the message is deleted locally. That allows you to recover deleted messages from the server if you need to. The Nokia phone's native email Inbox is always a mirror of the account on the server. The deleted messages are erased from the server immediately after the phone connects to the server the next time.

Handle attachments. Originally, email messages were just plain text. Later, Multipurpose Internet Mail Extensions (MIME) technology was developed to encode arbitrary binary files into text, and hence allowed the files to be sent as part of the email message (i.e., attachments). The Nokia native email client on Series 60 phones supports MIME attachments for both sending and receiving email messages.

The current generation of Nokia Series 40 phones (e.g., the Nokia 6230 camera phone) does not support attachments in email. You can still use MMS to send media files from the phone to an email address **[Hack #59]**. But you cannot receive files in email messages.

Figure 10-15 shows how to add an attachment from the phone's Gallery. Notice the difference in the header bar in the first and last screenshots (i.e., after the attachment is added the message size is larger, and a paper-clip icon indicates there is an attachment).

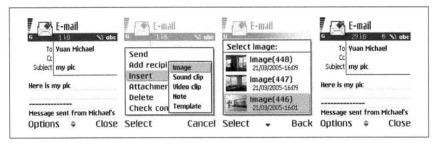

Figure 10-15. Adding an image as an attachment to an email message

You can open and save the message attachments, if there are any, from the Options menu (see Figure 10-16). If the phone recognizes the attachment file format as a supported media format, it provides the option to save the attachment file to the phone's Gallery.

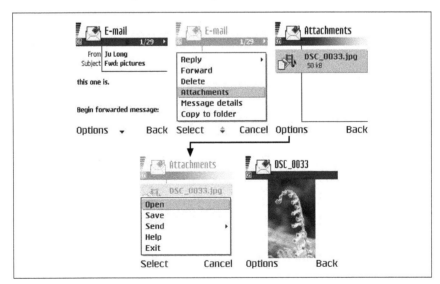

Figure 10-16. Opening a message attachment

If the attachment is in a format the phone does not support, the phone will not allow you to save it to the Gallery. As shown in Figure 10-17, there is no Save option for the PDF file attachment.

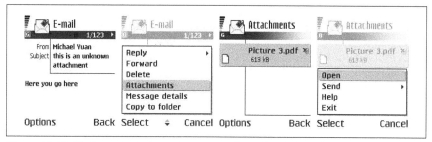

Figure 10-17. No Save option for an unsupported email attachment (PDF in this case)

If you install a PDF reader program **[Hack #39]** on your phone, the PDF reader will register *.pdf* files (and their appropriate MIME types) as "recognized" file formats. Then you will be able to open PDF attachments using the reader program from within the Messaging Inbox.

You can still send the attachment to another device or computer. Or, you can use a file browser, such as FExplorer **[Hack #20]**, to get the attachment file directly from the mail folders.

Use an HTML Email Client

In recent years, many email clients began supporting Hyper Text Markup Language (HTML) formatting tags in messages. An HTML message can have rich fonts, layout controls, hyperlinks, and inline images. The inline images can be included in the same message as attachments. While the Nokia native email client handles plain-text messages and attachments very well, it does not support HTML rendering or inline images. To view HTML email messages on your phone, you need third-party software.

When I have a choice, I always try to avoid sending HTML email messages, since they are not compatible with all email reader clients.

ProfiMail. ProfiMail is a Symbian-based email client application for Nokia Series 60 devices. It is developed by Lonely Cat Games and is available for sale and trial at *http://www.lonelycatgames.com/mobile/profimail/profimail. html*. The process to set up mailboxes in ProfiMail is very similar to the setup process for the native email client.

ProfiMail renders HTML markup in email messages. If you have a multipart message with several text and image components, ProfiMail renders the components in the order they appear in the message. So, you can have images between paragraphs of text even if you do not use HTML (see Figure 10-18).

A ProfiMail feature many people like is the ability to display messages in small fonts. That fits more message content in a single screen. For instance, ProfiMail splits the screen into a message header panel and a message body panel to ease message navigation (see Figure 10-18). However, the problem with small fonts is that they are very difficult for seniors or vision-impaired people to see [Hack #48].

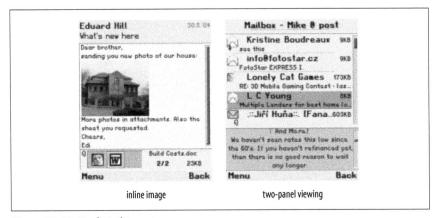

inline image   two-panel viewing

Figure 10-18. ProfiMail in action

ProfiMail also comes with a built-in file management program for the Symbian OS. It is very similar to the FExplorer file manager [Hack #20]. Using the file manager, you can save any email attachment (e.g., images, PDF documents, Word documents, etc.) directly to your phone memory or MultiMediaCard (MMC) card. Compared with the Nokia native email client, ProfiMail allows you to open more types of file attachments. For instance, you can open and browse zip files or the text in a Microsoft Word file.

Finally, ProfiMail supports rule-based filters that can automatically classify incoming messages based on their subject line, or on the text content in the body. You also can configure ProfiMail to automatically delete messages that it thinks are spam.

EmailViewer. ProfiMail is a "heavyweight" email client that works only on Series 60 phones. What if you need to display HTML messages and attachments on a smaller Series 40 phone? The EmailViewer program (*http://www. reqwireless.com/emailviewer.html*) is the answer. It renders HTML messages, as well as common document formats for email attachments, such as PDF files, zip files, and MS Word files. EmailViewer is a Java application, and it works on both Series 40 and Series 60 phones.

Under the hood, EmailViewer utilizes a proxy server to do all the heavy lifting. The proxy server fetches the message, renders it into a stream of bytes in a compact and proprietary format, and then delivers the bytes to the client. The mobile client simply displays the data in the stream according to a pre-defined convention. There is no need to parse, decode, render, and then lay out the HTML content on your mobile phone.

You have to install a proxy server licensed from Reqwireless, or subscribe to Reqwireless's proxy service, for the EmailViewer client to work.

EmailViewer and WebViewer **[Hack #50]** are developed by the same company, and they utilize the same approach for proxy-based content rendering.

Web-Based Email

If you use web-based email services, such as Yahoo! Mail, Hotmail, Gmail, and AOL Mail, you can check email directly from your phone's web browser. Most popular web-based email services offer WAP interfaces that work with almost any phone browser. The advantage of this approach is its simplicity. You do not need to configure the email servers on the phone. You do not even need a high-end data plan—a regular WAP browser will suffice in many cases.

Many ISPs or corporate email systems also provide web interfaces. But they are typically HTML-only. So, you need an HTML-compatible browser on the phone to access those services via the Web **[Hack #50]**.

However, the WAP interfaces do not support all advanced features of the web-based email service. Figure 10-19 shows how to check and retrieve messages from Yahoo! Mail's WAP service. HTML message formatting and attachments are not supported. But long messages are divided into multiple pages for easy reading on mobile phones.

Figure 10-20 shows how to compose and send an email message from a Yahoo! Mail account. Again, HTML formatting or file attachments are not supported. If you click the To or Cc link on the message composition form, you will be able to select recipient addresses from your Yahoo! address book.

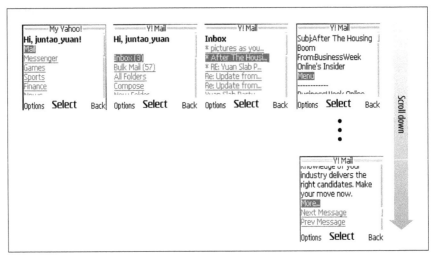

Figure 10-19. Getting email from Yahoo! Mail

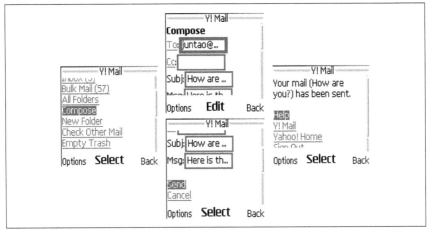

Figure 10-20. Sending email from Yahoo! Mail

 ## Manage Your Mobile Email

Make mobile email more efficient with email forwarding and filtering solutions.

Most of us have more than one email address these days. For example, I have separate email accounts for business, personal, news group, and web commerce needs. However, not all email accounts can be accessed directly from a Nokia mobile phone:

- Some accounts might require SSL or APOP authentication, which is not supported on all phones.

- Some accounts might require connections from computers in the same corporate network. The mobile phone connects from the wireless operator's network, and hence they are certainly outside of most corporate firewalls.

- Servers such as Microsoft Exchange are known to provide substandard POP/IMAP services. They prefer you to connect via their proprietary protocols, which are not supported on Nokia phones.

Even if you can access all your accounts from your phone, it is a hassle to thumb through the phone's lengthy menu options (especially for Series 40 devices) and check the mailboxes one by one. If you do not check all the mailboxes at one time, you might forget which ones you already checked the next time, and you'll have to go through all of them again. It would be nice to have a mobile-accessible email account that captures all your email messages in a single mailbox. In addition to providing convenient access to all your messages, an account dedicated to aggregating your email can make mobile email management a lot easier than before. Here are some use cases for a dedicated mobile email account:

Filtering

> We typically do not need to read all messages on our phone. For instance, one of my email accounts receives an alert message from every post to a discussion forum I monitor; another account is subscribed to a high-volume mailing list. I certainly do not want to read these messages while I am on the go. You can set up the aggregation account to aggregate only certain messages, and to filter out some messages for later processing when you are in front of a desktop computer.

Sandbox

> Multiple email clients from phones and desktop computers create message management issues. In particular, as I discussed earlier, if you delete a message from your phone, the desktop client cannot recover it or archive it later. You can use the mobile email aggregation account as an email "sandbox." You can download, view, and delete any message in the mobile account from your phone. The regular accounts still keep all your messages when you are home, where you can download the messages via your primary email client.

> When you send or reply to a message from your phone, it is a good idea to Cc yourself so that your desktop computer has a record of it later.

Ideally, the mobile email aggregation account should be a standard user account on a Unix/Linux server, with basic support for POP and SMTP. Most Unix/Linux-based ISPs or web hosts provide such services.

Server Forwarding

The easiest way to aggregate messages to a mobile email account is to have the original accounts forward the messages. If your original email account is on a Unix server, you can simply place a .*forward* file in your home directory with the following line (it keeps a copy of the message in the original account and forwards a copy to your mobile account):

```
/username, you@mobile.address.com
```

If you use the *procmail* utility (see *http://www.procmail.org/*) to manage your Unix email Inbox, you can add the following rule to the end of the *.procmailrc* file to copy all incoming messages to the *you@mobile.address.com* address:

```
:0c
*
! you@mobile.address.com
```

The advantage of *procmail* is that it supports a complex set of filtering rules. You can determine which messages to forward to the mobile account based on the sender and the content. For instance, you can avoid forwarding mailing list messages or autogenerated messages from forums. When you use it together with a spam filter, you can also avoid forwarding any message that might be spam.

For corporate email accounts (or non-Unix accounts) where you do not have access to the home directory, you should ask the administrator to set up the forwarding service.

Client Forwarding

Alternatively, you can forward messages from your desktop computer rather than the servers. The advantages are that the desktop computer is already set up to connect to all your accounts (i.e., no firewall issues), and you have control over what software to run on the desktop computer.

If your client computer is Unix/Linux based (including Mac OS X) or runs Cygwin on Windows, you can use the *fetchmail* program (*http://www.catb.org/~esr/fetchmail/*) to retrieve messages from multiple servers, filter those messages based on their headers, and then forward filtered messages to your mobile email account. The following entry in the *.fetchmailrc* file in your home directory instructs *fetchmail* to periodically retrieve messages from the *myregular@regular.address.com* account and to forward all messages to the dedicated mobile email account at *mymobile@mobile.address.com*:

```
poll pop.regular.address.com
        protocol pop3
        username mymobile here is myregular there
        password 'you cannot crack it'
        smtphost smtp.mobile.address.com
        fetchall
```

If your mobile email account is hosted on a Unix server and the ISP provides *fetchmail*, you can simply retrieve messages from your other accounts
and forward to the local inbox (i.e., *smtphost* or *localhost*). This is more reliable than remote forwarding.

If you use a Windows desktop computer and do not want to mess with Cygwin (*http://www.cygwin.com/*), you can use one of the several mail-
forwarding programs available for Windows. For instance, Mail Forward for
Windows (*http://www.sspi-software.com/mailfwd_win.html*) is a graphical
user interface (GUI)–based program for retrieving and forwarding messages. You can configure it to run periodically on your Windows box.

Use a Portal Server

You also can configure popular web mail services (e.g., Yahoo! Mail) and
your wireless operator portal to aggregate multiple email accounts.
Figure 10-21 shows the configuration page for aggregating multiple third-
party email accounts in Yahoo! Mail's Options menu.

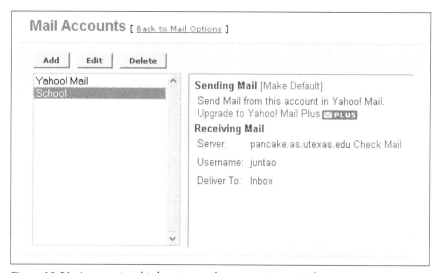

Figure 10-21. Aggregating third-party email accounts in your Yahoo! account

As I have discussed, those services can take remote messages and present
them in Wireless Markup Language (WML) format suitable for any phone.

A Mobile Gmail Gateway

#62   Access your Google Gmail account via a custom proxy.

Google Gmail is among the most popular web-based email services available today. The Google Gmail service not only offers 2 GB of free storage space, but it also provides great email management functionality such as filters and advanced message search. Although Gmail does not provide a mobile web site, it does support SSL access to the SMTP and IMAP servers. Hence, you can use techniques mentioned in "Send and Receive Email on Your Phone" **[Hack #60]** to make Gmail available on your phone.

In this hack, I will introduce an alternative technique by building a custom mobile portal for Gmail. It supports Gmail-specific features such as email filters, and gives you more control of your email to make better use of the limited wireless bandwidth. You can potentially use the same hack to aggregate your multiple email accounts and eliminate your dependency on operator or third-party WAP email portal servers.

> You can access the Gmail web site via an advanced HTML and JavaScript browser, such as Opera on Nokia Series 60 phones **[Hack #50]**. But direct web access is much slower than the proxy solution, and it does not work on Series 40 phones.

gmail-mobile (*http://sourceforge.net/projects/gmail-mobile*; GNU public license) is a PHP (*http://www.php.net*) application that sits on your web site, between your mobile phone's WAP browser and Gmail, brokering requests on your behalf and returning a mobile-appropriate view of your Gmail mail. You can catch a quick status update, read, and even reply to your Gmail—and more features are promised.

Install the Software

Download gmail-mobile (*http://sourceforge.net/projects/gmail-mobile*) and unpack the distribution (0.11 at the time of this writing, but your's is sure to be a later version) somewhere under your web server's document root, where the rest of your web site lives (ask your system administrator or service provider if you're not sure where this is):

```
$ tar -xvzf gmail-mobile-0.11.tar.gz
gmail-mobile-0.11/
gmail-mobile-0.11/AUTHORS
gmail-mobile-0.11/COPYING
gmail-mobile-0.11/INSTALL
gmail-mobile-0.11/README
```

```
gmail-mobile-0.11/TODO
gmail-mobile-0.11/compose.php
gmail-mobile-0.11/config.php
gmail-mobile-0.11/index.php
gmail-mobile-0.11/libgmailer.php
gmail-mobile-0.11/logout.php
gmail-mobile-0.11/main.php
gmail-mobile-0.11/star.gif
$ mv gmail-mobile-0.11 gmail-mobile
```

 That last bit of code renamed the *gmail-mobile-0.11* directory to something a little easier to type on my mobile
phone's keypad.

By default, gmail-mobile uses browser cookies to maintain state between
requests to Gmail's servers. If you have PHP Session (*http://www.php.net/
session*) installed, you can use it instead of cookies. Just comment out the
appropriate line in the *config.php* file in your newly unpacked *gmail-mobile*
directory. Here, I left things as they were, using the default cookie setting:

```
<?php

    require_once("libgmailer.php");

    /** Session handling method.
        You must at least choose (uncomment) one. **/

    /**** have PHP Session installed,
          prefer to use cookie to store session **/
    // $config_session = (GM_USE_PHPSESSION | GM_USE_COOKIE);
    /**** have PHP Session installed,
          prefer NOT to use cookie **/
    // $config_session = (GM_USE_PHPSESSION | !GM_USE_COOKIE);
    /**** do not have PHP Session installed **/
    $config_session = (!GM_USE_PHPSESSION | GM_USE_COOKIE);

?>
```

Run the Program

With the easy part out of the way (isn't it wonderful when installation and
configuration are the easy part?), you're ready to break out your mobile
phone's browser and muddle through typing on that minute keypad.

Before trying this out from your mobile phone (and to remove one variable
in case something doesn't work as expected), point your computer's web
browser to a URL corresponding to the *gmail-mobile* directory on your web
site—e.g., *http://www.example.com/~rael/gmail-mobile*.

You might actually need to tack /index.php onto that URL, but most PHP-enabled servers know to look for and serve up *index.php* as a default when no filename is specified and there's no static *index.html* in sight. The gmail-mobile package age includes just such an *index.php* file.

Your browser will respond in one of two ways. Either it'll serve up the raw WML source delivered by *gmail-mobile*, as shown in Figure 10-22, or it'll throw up its hands in confusion and prompt you to save the source as a file on your hard drive. If the source (displayed in your browser or saved and opened using something such as TextEdit on Mac OS X or Notepad on Windows) looks something like Figure 10-22 and doesn't seem to report any PHP or other errors, you're ready to switch to your mobile phone.

```
http://jane.local/gmail-mobile/

◀ ▶  C  +  ⊕ http://jane.local/gmail-mobile/          Q▾ Google

<?xml version="1.0" encoding="utf-8"?>
<!DOCTYPE wml PUBLIC "-//WAPFORUM//DTD WML 1.1//EN" "http://www.wapforum.org/DTD/wml_1.1.xml">

<wml>
  <card title="gmail-mobile">
    <p><b></b></p>
    <p>login: <input name="login" type="text" value="" /></p>
    <p>password: <input name="pass" type="password" /></p>
    <p>timezone: GMT <input value="+8" name="tz" type="text" size="4" />:00</p>
    <p>
      <do type="accept">
        <go method="POST" href="main.php">
          <postfield name="login" value="$(login)" />
          <postfield name="sum" value="1" />
          <postfield name="pass" value="$(pass)" />
          <postfield name="tz" value="$(tz)" />
        </go>
      </do>
    </p>
    <p align="right">
      <do type="prev" label="[&lt;&lt;]">
        <prev />
      </do>

    </p>
  </card>
</wml>
```

Figure 10-22. Raw gmail-mobile WML page

Launch your mobile phone's WAP browser and key in the appropriate URL to reach the *gmail-mobile* directory on your web site, as explained earlier. After a few moments of churning (WAP is lightweight, but most mobile bandwidth is on the light side too), you should be greeted with a login screen (Figure 10-23, left). Key in your Gmail login (*username@gmail.com*) and password, alter the time zone if you feel so inclined, and click OK. Just where you find the OK button will vary from phone to phone and from WAP browser to WAP browser. In most cases, you can just click the center key on the navigation pad. I also found it under the Options → Service options → OK menu on my Nokia Series 60 phone, as shown in Figure 10-23, right.

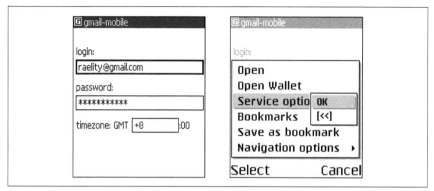

Figure 10-23. Logging into Gmail from your mobile phone WML browser

A few more moments of churning and you should see a summary view of your Gmail account (Figure 10-24, left). To visit any of the folders, navigate over the appropriate link and select it, much as you would links in a regular browser—albeit with esoteric keystrokes rather than a mouse. Figure 10-24, right, shows my rather empty Inbox.

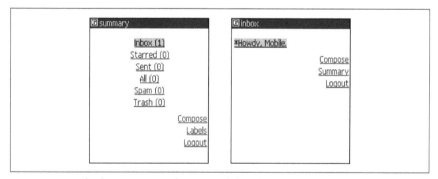

Figure 10-24. Checking your Gmail status and the Inbox

You can visit any message (Figure 10-25, left, shows a sample email message) in any of your mailboxes by selecting its link. Compose a new message by selecting the Compose link; reply using the Reply link at the bottom of a message. Figure 10-25, right, shows the composition window in action.

While you can't (at least at the time of this writing) create, alter, or delete Gmail labels, you can see what they are (Figure 10-26, left) by following the Labels link on the Summary screen (Figure 10-26, right, shows all of my messages labeled Peeps).

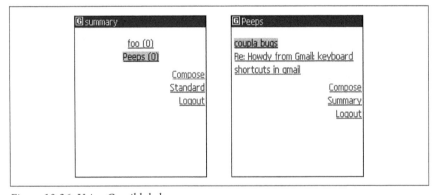

Figure 10-25. Reading and responding to Gmail messages

Figure 10-26. Using Gmail labels

The WAP-based gmail-mobile might not be the spiffy, tricked-out Gmail interface that you've come to expect, but it's a great way to take your Gmail with you.

This hack is adapted from "Gmail on the Go" **[Hack #77]** in *Google Hacks*, Second Edition (O'Reilly, 2005) by Tara Calishain and Rael Dornfest.

—Rael Dornfest

SMS from a Computer
HACK #63

Send SMS messages with a real keyboard!

SMS is often the best way to reach people on the move. Since SMS messages are pushed to the phone over the cellular network, it can be much faster than regular email. However, it is inconvenient to type and send SMS messages on a phone keypad **[Hack #58]**. In this hack, you will learn how to type and send SMS messages from a regular keyboard on your desktop computer.

The key challenge I need to address in this hack is how to connect a computer to the wireless network where SMS messages are transmitted. You can do this in three ways:

- You can purchase and install a Global System for Mobile communications (GSM)/GPRS PC card (e.g., GlobeTrotter, Ubinetics, or Merlin) and directly connect the computer to the wireless network. Most new cards are expensive, but you can buy a used Merlin G100 card for as little as $50. You need to put your SIM card into the PC card—you cannot use your phone while you are SMS messaging from the PC. Due to their price and limitations, those PC cards are not the focus of this hack.

- You can use a connected mobile phone as a relay, with the phone connected to the PC via Bluetooth, IR, or cable links. Once you've set this up, your PC is directly connected to the wireless network and can relay messages from the computer to the network.

- Some remote services provided by the wireless operator or third-party portals can forward the messages submitted via the Internet to the wireless network.

Now, let's check out exactly how these work.

Use a Windows Computer

On a Windows computer, the Nokia PC Suite bundles an SMS program for sending messages over a connected phone (see Figure 10-27).

Alternatively, you can use the DesktopMessage program from Mobile Ways (*http://mobileways.de/M/1/1/0/*) to send SMS messages via a Series 60 device. In both cases, the phone is used as a message relay, as described earlier.

Use a Mac Computer

Mac OS X's Address Book application has built-in support for sending SMS messages via Bluetooth-paired Nokia phones. First, you should make sure the device is connected to the Address Book. If it isn't, just press the Bluetooth icon on the Address Book's main window and confirm the connection request on the phone.

Now, Ctrl-click (or right-click) any phone number field to bring up the context menu. Select the SMS Message menu item to send a message to this phone number (Figure 10-28). On the recipient phone, the SMS message appears to come from the phone paired with your Mac.

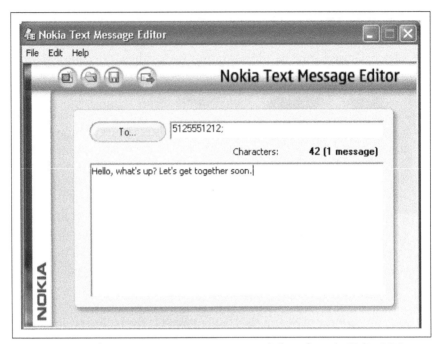

Figure 10-27. Sending an SMS message via a connected phone from the Nokia PC Suite

Use a Web Portal

Using any Internet-ready computer, you can send SMS messages to almost any phone via web portals. Each wireless operator has its own SMS portal page. Third-party mobile portals, such as Yahoo! Mobile (*http://mobile. yahoo.com/*), provide operator-independent SMS sending services (see Figure 10-29).

> You can also send an SMS message to a phone via the operator's email gateway **[Hack #59]**. The email gateway makes it easy for you to send SMS messages from your own web site. Just write a CGI script to send the email!

Send Lots of Messages

The SMS sending methods I discussed so far are good for occasional messaging by hand. They are not suited for high-volume, automated message-sending campaigns. Unless you have an unlimited SMS service plan, each message you send over your relay phone is charged the regular retail rate for SMS (typically 10 cents per message in U.S.). That makes this a very expensive method if you need to send many messages. While you can automate

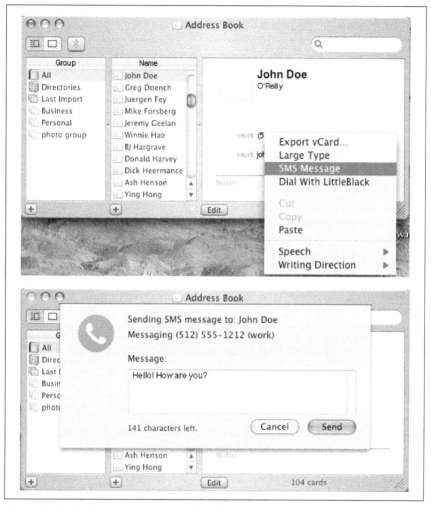

Figure 10-28. Sending an SMS message from the Mac Address Book using a Bluetooth-connected phone as a relay

message posting on free web portal forms or gateway email addresses, doing so is a violation of the service terms. Those SMS portal services are good at detecting and blacklisting SMS robots. To send large volumes of SMS messages to many users automatically and economically, you need to purchase additional software and services.

The MercuryXMS (*http://www.businesms.com/*) and Simplewire (*http://www.simplewire.com/*) software packages come with subscription-based gateway services that forward your SMS/MMS/WAP messages to operators in many countries. Both packages allow you to write applications

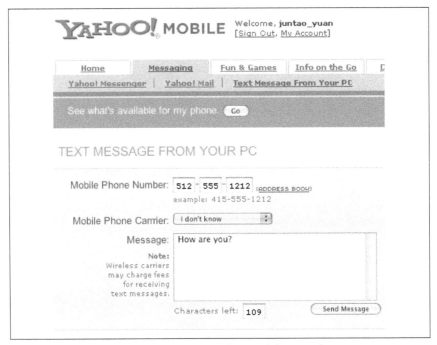

Figure 10-29. Sending an SMS message via Yahoo! Mobile's portal page

using real programming languages (e.g., Java, .NET, Perl, and other scripting languages) to automate the message-sending process. Your application can also receive messages sent to special phone numbers if you set up the receiving service.

HACK #64 Make Use of Alert Services

Alert yourself of events, news, and price changes on the go.

Most mobile portal web sites, including both wireless operator sites and third-party sites, offer SMS alert services to deliver live alerts and updates to mobile users. Those services are typically free of charge. But of course, you need to pay for the SMS service subscription or pay a per-message fee to receive those alerts. In this hack, I will use the Yahoo! Mobile portal and the T-Mobile portal as examples to show what you can do with these alert services.

> In addition to SMS-based alert messages, mobile portals, such as Yahoo!, can also deliver alerts via regular email or Internet IM.

Figure 10-30 shows the alert categories available on the Yahoo! Mobile portal site. You can click any of the categories and schedule alerts.

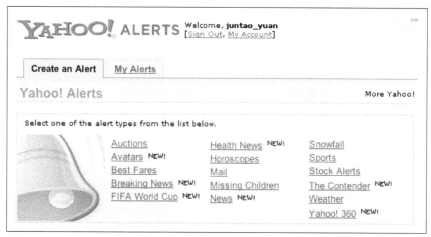

Figure 10-30. Alert categories available from Yahoo!

Figure 10-31 shows my alerts. The "Deliver to" icons indicate whether the alerts are delivered to email, Yahoo! Instant Messenger, or a mobile phone.

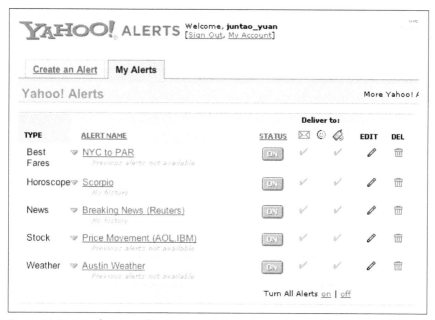

Figure 10-31. My alerts in Yahoo!

Alert Services

The portal can aggregate and monitor information from the Internet. The alert service sends SMS messages to your phone at certain times of the day or when the monitored parameters hit a certain threshold. Some of the most commonly used alert services are as follows:

News
> You can select a news source (e.g., AP or Reuters) and deliver all the breaking-news headlines to your phone whenever they become available.

Auction tracking
> You can track bids in auctions and have an alert sent to you when a new bid meeting a certain price level is posted. The Yahoo! auction tracker service integrates with the Yahoo! auction marketplace.

Airfare tracking
> You can specify a list of travel routes (i.e., pairs of cities) and have the portal monitor their prices. If the price drops by a certain percentage (e.g., 25%), the portal sends you an alert. On Yahoo!'s mobile portal, monitoring is integrated with Yahoo!'s Travelocity service.

Stock alerts
> The portal can monitor the price of your stocks for you. When the price increases or decreases beyond some preset criterion (e.g., a fixed price point or percentage change), an SMS alert message is generated and is sent to your phone.

Daily weather report
> You can specify a list of cities and have their weather forecasts delivered as SMS messages at certain times of the day.

Your Own Alerts

Some portal sites (e.g., the T-Mobile portal) allow you to set up your own alerts. You can configure the portal to send SMS messages to yourself at a future date and time. This is a great way to remind yourself of anniversaries or appointments.

> Of course, another alternative is to enter appointments and anniversaries into the Calendar on your phone. You can associate an alarm tone for any calendar event.

Search Google Via SMS

#65 Use Google and the Internet as the ultimate SMS information services, for free.

As suggested recently in a *New York Times* article (see "All Thumbs, Without the Stigma"; visit *http://tech2.nytimes.com/mem/technology/techreview.html?res=9E00E6DE-163FF931A2575BC0A9629C8B63*), the thumb is the power digit. While the thumb-board of choice for executives tends to be the BlackBerry mobile email device (*http://www.blackberry.com/*), for the rest of the world (and many of the kids in your neighborhood) it's the cell phone and Short Message Service (SMS).

SMS messages are quick-and-dirty text messages (think mobile instant messaging) tapped into a cell phone and sent over the airwaves to another cell phone for around 5 to 10 cents apiece.

But SMS isn't just for person-to-person messaging. In the UK, BBC Radio provides so-called shortcodes (really just short telephone numbers) to which you can SMS your requests to the DJ's automated request-tracking system. You can SMS bus and rail systems for travel schedules. Your airline will "SMS you" updates on the status of your flight. And now you can talk to Google via SMS as well.

Google SMS (*http://www.google.com/sms/*) provides an SMS gateway for querying the Google web index, looking up phone numbers, seeking definitions, and doing comparative shopping in the Froogle product catalog service (*http://froogle.google.com/*).

Search Options

Simply send an SMS message to U.S. shortcode 46645 (read: GOOGL), as shown in Figure 10-32, with one of the following forms of query. You'll receive your results as one or more SMS messages labeled, for instance, "1 of 3," "2 of 3," etc. Notice that the responses contain no URLs or links. What's the point of having these when you can't click them?

> While the sender usually bears the cost of sending an SMS message (typically between 5 and 10 cents apiece), automated messages such as those sent by Google SMS are usually charged to you, the receiver. Unless you have an unlimited SMS plan, all that Googling can add up. Be sure to check what's included in your mobile plan, check your phone bill, or call your mobile operator before you spend a lot of time (and money) on this service.

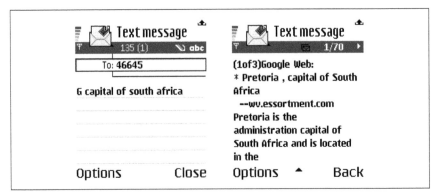

Figure 10-32. Searching Google via SMS

Web search. Search the Google Web by prefixing your query with a G (upper- or lowercase). You'll receive the top two results in return, formatted as text snippets, hopefully containing some useful information:

```
g capital of south africa
G answer to life the universe and everything
```

Local business listing. Consult Google Local's business listings by passing it a business name or type and city, state combination, or Zip Code:

```
vegetarian restaurant Jackson MS
southern cooking 95472
scooters.New York NY
```

 The Google SMS documentation suggests using a period (.) between your query and city name or Zip Code, to be sure that you're triggering a Google Local Search.

Residential phone number. Find a residential phone number with some combination of first or last name, city, state, Zip Code, or area code:

```
augustus gloop Chicago il
violet beauregard 95472
mike teevee ny
```

 As with any Google Phonebook query, you'll find only listed numbers in your results.

Froogle price. Check the current prices of items for sale online through Froogle. To trigger a Froogle lookup, prefix your query with an F (upper- or lowercase), the word *price*, or the word *prices* (the latter two will also work at the end of the query):

```
f nokia 6230 cellphone
price bmw 2002
ugg boots prices
```

Definition. Instead of scratching your head, trying to understand just what Ms. Austen meant by disapprobation, ask Google for a definition. Prefix the word or phrase of interest with a D (upper- or lowercase) or the word *define*:

```
D disapprobation
define osteichthyes
```

Calculation. Perform feats of calculation and conversion using the Google Calculator:

```
(2*2)+3
12 ounces in grams
```

Zip Code. Pass Google SMS a U.S. Zip Code to find out where it's used in the country:

```
95472
```

Google SMS is sure to sport more features by the time you read this. Be sure to consult the "How to Use Google SMS" page at *http://www.google.com/sms/howtouse.html* for the latest information, or—for the real thumb jockeys among you—subscribe your email address to an announcement list from the Google SMS home page.

Use a Rich Mobile Client

While the Google SMS is easy to use, the SMS client user interface (UI) is not optimized for query and response types of operations. The prefix and suffix of the query strings in different modes can also be hard to remember. If you frequently search for local information, it is also a hassle to type in your address repeatedly for each search.

Erik C. Thauvin developed a Java application called GooglME to wrap around Google SMS features. It works on most Series 40 and Series 60 phones. You can download GooglME from *http://mobile.thauvin.net/j2me/*. GooglME automatically adds prefixes to queries based on the search type you select (see Figure 10-33). For instance, if the search type is *Froogle*, GooglME adds the letter "F" and a space in front of the query string in the text box, and then sends the SMS query over to Google. GooglME also automatically appends your address (entered via the Options → Settings menu) at the end of every query in the Local search types.

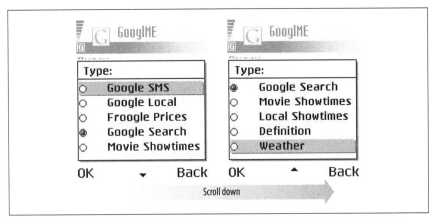

Figure 10-33. Selecting the search type in GooglME

GooglME's UI is simple and elegant. Figure 10-34 shows the query and results of a regular Google search performed over GooglME.

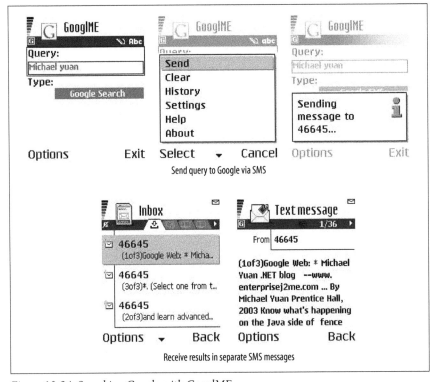

Figure 10-34. Searching Google with GooglME

This hack is adapted from "Google on the Go" [Hack #67] in *Google Hacks*, Second Edition (O'Reilly, 2005) by Tara Calishain and Rael Dornfest.

—Rael Dornfest

HACK #66 Mobile Instant Messaging

Send and receive instant messages (IMs) to and from popular IM clients from your phone.

For many Internet users, instant messages (IMs) have become a part of daily life. They are faster than email but less intrusive than phone calls—perfectly suited for short, conversational-style communication between two or more parties. Some IM systems support file attachments or even audio and video streaming between chat parties. All these features are typically available free of charge. As a result, instant messaging is now widely used in both personal and corporate communications. Here is a list of the most popular IM systems:

- Microsoft Messenger (MSN; *http://messenger.msn.com/*)
- AOL Instant Messenger (AIM; *http://www.aim.com/*)
- Yahoo! Messenger (YIM; *http://messenger.yahoo.com/*)
- ICQ (*http://www.icq.com/*)
- Jabber (*http://www.jabber.org/*)
- Internet Relay Chat (IRC; *http://www.irchelp.org/*)

A very annoying problem in the IM world is the fact that popular IM systems do not interoperate. For instance, you cannot send a message from a Yahoo! client to an AOL client.

The open source Gaim project (*http://gaim.sourceforge.net/*) develops a third-party, alternative IM client that understands multiple IM protocols including MSN, AIM, YIM, and ICQ. Gaim runs under Linux, Mac OS X, and Windows operating systems. But Gaim does not work on smartphones.

Some Jabber IM servers run "gateways" for other IM systems, and hence give their clients access to those non-Jabber IM systems.

To make the smartphone a true "personal communications center" or "mobile office," it has to support popular IM systems. In this hack, you will learn ways to access Internet IM messages from Nokia phones.

Series 40 Native Client for AIM

In all popular IM systems, AIM has the best support for mobile IM. If you purchase a Nokia Series 40 phone (e.g., the Nokia 6800/6820 messaging phone with folded keyboard) from a U.S. operator as part of your service contract, the phone is probably preloaded with AIM's chat software. To check, just go to the phone's Main menu and scroll through it to see whether there is an IM application. If there is, the phone is configured to access AIM services.

> In fact, the AIM client on a Nokia Series 40 phone is a Wireless Village client configured for AIM services. Please see the section titled "The Future of Mobile IM," later in this hack, for more information on the Wireless Village system and protocol.

After you select the IM menu, type in your AIM account credentials to connect to the AIM server. Then you can send/receive messages and manage your buddy list.

If your phone does not bundle AIM software, do not panic! Read on to find out other ways to access AIM and other IM systems from any Nokia Series 40 and Series 60 phone.

Wireless IRC

The IRC protocol is not controlled by any commercial company, and hence anyone can develop IRC clients. Several IRC clients are available for mobile phones.

The WLIRC project (*http://wirelessirc.sourceforge.net/*) develops a J2ME-based open source IRC client for mobile phones. It runs on almost any Nokia Series 40 and Series 60 phone. The WLIRC client connects to IRC servers through the standard IRC protocol via TCP/IP sockets. If your phone or network does not support raw TCP/IP connections, you can run an HTTP gateway included in the WLIRC software download. Your phone connects to the HTTP gateway and then to IRC servers on the Internet.

For Nokia Series 60 phones, the WirelessIRC program from Mobile Ways (*http://mobileways.de/M/1/4/0/*) supports chatting with other IRC users. The WirelessIRC program has a nice UI and supports file exchange over IRC channels.

Instant Messaging Via SMS

Mobile phones have long had their own "instant messaging" protocols: SMS and, more recently, MMS. SMS messages are not exactly "instant," since SMS is low-priority traffic in the mobile network and, in theory, a message can take hours to deliver. But in reality, most SMS messages are delivered within seconds. So, SMS is a reasonable option for delivering IM messages.

> While different mobile network operators have different SMS networks, they interoperate well.

The recent versions of the AIM and YIM clients allow you to use mobile phone numbers as message destinations. When you send a message to a mobile phone number (prefixed with a + and the country code, as in +18885551234), the phone receives an SMS message containing the sender's ID and the message content. The SMS message appears to come from a service phone number set up by the operator (e.g., very short or very long numbers that are obviously not regular phone numbers). If you want to reply to that message, you can simply reply to the SMS message. Make sure you do not change the recipient phone number (i.e., the service number where the incoming message is sent from) or the message subject in the reply message. The reply message is received by the IM service listening at the service number and is then forwarded to the IM client of the original sender.

> Since IM providers need to set up special service phone numbers to deliver messages and receive replies via SMS, they have to enter into business agreements with wireless operators. For example, as of April 2005, YIM works with Cingular, AT&T, Sprint, and Verizon in the U.S. You need to check with your IM provider for an updated operators list for your area.

As an added bonus, you can use the same SMS client to interoperate with all IM services that support IM-over-SMS. From the SMS client's point of view, the different IM services are merely identified as different service phone numbers.

In the case of YIM, you can also send command messages to a special service number, 92466 (the number representation for "YAHOO"), to block certain Yahoo! users from sending IM messages directly to your phone. Available commands are as follows:

- To block a YIM user, the command is *block YahooID*.
- To block everybody, the command is *block all*.

- To unblock a YIM user, the command is *unblock YahooID*.
- To unblock everybody, the command is *unblock all*.

While the IM-over-SMS solutions are easy to use, they do not replace real IM clients. For instance, SMS doesn't support buddy lists or a presence status (e.g., online or offline). In addition, not all IM providers and/or wireless operators support IM-over-SMS. So, let's explore some "real" IM client solutions on mobile phones.

> *Presence* is a short status message often used in IM applications. For instance, your presence in an IM system can be "work," "home," "busy," "play," etc.

Agile Messenger

If you use a Nokia Series 60 phone, Agile Messenger is an excellent Symbian-based IM client that handles multiple IM protocols (e.g., ICQ, MSN, YIM, and AIM). You can think of Agile Messenger as a closed source equivalent to Gaim on Nokia Series 60 phones. You can download and install the Agile Messenger software from its web site at *http://www. agilemobile.com/downloads.html*. Make sure you get the correct version for your device.

Once you start the Agile Messenger program, it shows you a list of IM services. You can add an existing account for any of these services (see Figure 10-35). Then Agile Messenger connects to the selected service, asking you to choose your GPRS connection profile along the way. You need to make sure the GPRS connection you select supports all TCP/IP ports **[Hack #4]**.

> The Agile Mobile Messenger client uses TCP/IP to connect to IM servers directly. There is a risk that the wireless operator (especially in the U.S.) might use network filters to block those TCP/IP ports. So, try it out before you buy!

Your buddy lists should appear immediately after you connect. You can also add contacts through the menu. The colors of the contacts in the list indicate the network the contacts are subscribed to: red is for YIM; yellow is for AIM; blue is for MSN; and so on. If you see small icons next to the contacts, their online status is either "offline" (indicated by a small white cross in a red circle) or "busy" (a small white clock symbol). Figure 10-36 shows how to select a contact and start chatting.

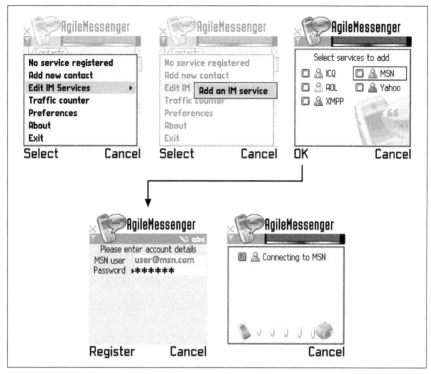

Figure 10-35. Adding an IM account to Agile Messenger

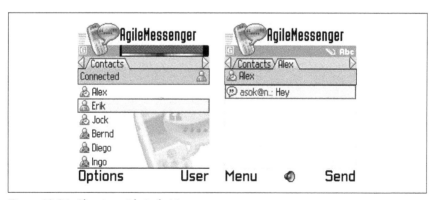

Figure 10-36. Chatting with Agile Messenger

After you finish a chat session, you can leave Agile Messenger running in the background by pressing the red End button to return to the Main menu. The background-running Agile Messenger will notify you of new messages, even when you are in other applications.

TipicME

If you do not own a Nokia Series 60 phone, or if your wireless plan does not provide an unrestricted GPRS access point, you are out of luck in terms of Agile Messenger. But TipicME (*http://www.tipic.com/tipicme*) might be able to help you here. TipicME is a J2ME-based IM client that works on both Nokia Series 40 and Series 60 devices. It only requires a data plan that supports HTTP, the protocol used for browsing the Web. TipicME works with all popular IM services, including ICQ, MSN, AIM, YIM, and Jabber/XMPP. It supports all popular IM client features, such as buddy lists and presence.

Under the hood, the TipicME client makes HTTP connections to a proxy server hosted by Tipic, which in turn connects to multiple IM servers. So, to use TipicME, you have to set up an account with the Tipic proxy server and subscribe to its services first.

Since HTTP connections are pull based, the TipicME client must poll the proxy server periodically for new messages and presence status updates. The more frequently the polling occurs, the more accurate the message update is on the client. However, frequent polling increases your bandwidth usage.

Instant Messaging Via a Portal

For occasional mobile IM users, MSN, AIM, and YIM provide WAP web portals to their IM services. Figure 10-37 shows how to send a YIM message on the Yahoo! portal. It allows you to choose a recipient from your online address book (buddy list) and tells you whether the contact is online. Since the web client is HTTP based, you have to poll the server manually to check for new messages.

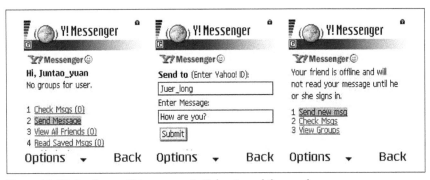

Figure 10-37. Sending an IM message via Yahoo!'s mobile portal

To reduce the amount of typing you have to do on the phone keypad, the Yahoo! portal allows you to choose from some predefined common

sentences to assemble your message (see Figure 10-38). This is a nifty feature that is commonly available in rich client messaging clients.

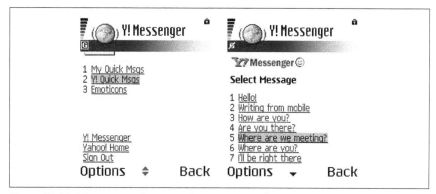

Figure 10-38. Message templates at Yahoo!'s IM portal

 Some mobile portal services (e.g., the portal from your network operator) might provide portals that interoperate with
all IM services. Check your operator's web site for availability. For more on mobile web portals, see "The Mobile Web
in Bite-Size Chunks" **[Hack #51]**.

The Future of Mobile IM

The future of mobile IM on Nokia phones, obviously, is for Nokia to integrate an IM client directly into the Messaging application on the phone.
Actually, some new Nokia models (both Series 40 and Series 60 devices)
come with a client for a messaging system known as the Wireless Village.
Wireless Village–compatible phones come with a Chat application under
the Messages (or Messaging) Main menu (see Figure 10-39).

Figure 10-39. Wireless Village Chat client in a Nokia 6230 phone

Wireless Village is developed by Motorola, Sony-Ericsson, and Nokia. It is
an Open Mobile Alliance (OMA) standard and is supported by all major
wireless operators. The idea is that each operator can run its own messaging

servers and offer branded IM service to their own customers. Then, the operators can interoperate and provide mobile IM services to all mobile subscribers—like how SMS systems interoperate today.

Wireless Village uses it own underlying messaging protocol known as the Mobile Instant Messaging and Presence Services (IMPS). The IMPS protocol is an XML-based protocol that supports push-based messaging. Push messaging allows users to be notified of incoming messages even when the IM client is not running.

> IMPS is similar to another IETF standard known as the *Extensible Messaging and Presence Protocol* (XMPP). XMPP is the underlying protocol for the popular Jabber IM system.

You can use the IMPS presence service in other mobile phone applications beyond IM. For instance, you can use presence to indicate whether it is a good time for others to call you. The Contacts application in new Nokia phones integrates with the presence service. It allows you to update your own presence status as well as to check other people's status (see Figure 10-40). You can configure the presence status to change automatically with profile changes. For instance, you can associate the Meeting profile with the Busy presence status so that people know to avoid calling you while you are in a meeting.

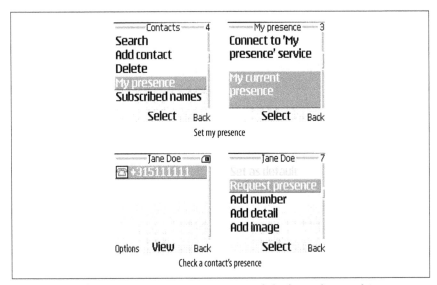

Figure 10-40. Changing your own presence status and checking other people's status

However, it is not clear whether Wireless Village will be a commercially successful IM system. Currently, the IMPS chat and presence features on Nokia phones are not supported by most operators. While Wireless Village enjoys wide support within the wireless industry, it will be useful only after the incumbent desktop IM services buy in. MSN, AIM, and YIM have so far demonstrated deep skepticism toward any third-party application that might circumvent their proprietary IM protocols. Time will tell whether IM standardization efforts from mobile phone vendors will be successful.

—Frank Koehntopp and Michael Yuan

Mobile Multimedia
Hacks 67–75

Nokia Series 40 and Series 60 smartphones are very powerful personal multimedia devices. For when you need to create content, the camera and audio recorder on the phone enable you to capture photos, video clips, and audio clips on the go. Then you can share your media instantaneously with friends around the world. However, the mobile phone photos and video clips often are of low quality. In this chapter, you will learn how to edit and enhance the media you capture on your phone and make them look more professional. Then you can share them with a wider audience.

For when you want to listen to or view media, a mobile phone has the perfect form factor to become a personal multimedia player. Using hacks covered in this chapter, you will be able to watch DVD movies and listen to MP3 music on your phone, anywhere and at any time.

HACK #67 Enhance Mobile Phone Photos
Get the best possible photos from your tiny phone camera.

Camera phones have become increasingly popular. Most new Nokia phone models are equipped with cameras. However, as more and more people use camera phones, the quality of camera photos has emerged as a major issue for users. In this hack, you will learn how to take better photos and how to make your mobile phone photos more appealing using photo-editing software.

When using conventional digital cameras, you can control a lot of factors, such as exposure, focus, and resolution at the moment when you take the picture. However, when using camera phones, you do not have that luxury. Camera phones typically have only one control: the one that lets you take the picture. Now, take a look at your camera phone. Does that tiny plastic-like lens look like something that can take a sophisticated picture? Probably not. You trade a lot of the basic controls that come with your conventional camera for the mobility you get with your camera phone. To take better pictures with your less-than-perfect camera phone, you need to enhance the pictures *after* you take them. Before you do that, you need to know what you can expect from your camera phone.

Here are the components that make up a camera phone, and their limitations:

Image sensor
> As with other digital cameras, a camera phone uses either a CCD or a CMOS sensor (the two main types) as its image sensor. Both sensors work by converting the light entering the lens into an electrical signal that is processed to produce the photograph. Sensors used on camera phones usually do not have the same quality as those in conventional digital cameras. As a result, camera phone photos are not as sharp and vivid as the ones taken with conventional cameras.

Lens
> Like the image sensor, the lens in a camera phone is usually of limited quality too. It is small and has a fixed focus and aperture. This cannot compete with the lenses on conventional digital cameras. With a fixed focus, aperture, and shutter speed, you cannot adjust the exposure time. You cannot focus on the objects off the center of your picture frame, either. Some lenses on camera phones can suffer from a problem known as *barrel distortion*, whereby lines at the edges of the picture seem to be bowed like a barrel.

Resolution
> The quality of the photos ultimately depends on the resolution of the sensor. The standard resolution for most camera phones on the market is 640×480. That translates to a mere 0.3 megapixels, which indicates the number of *pixels*, or individual dots of light, that make up the image. This is far inferior to a conventional camera's 2–5 megapixels. Poor resolution is obvious when you notice the "graininess" of the photos. To produce acceptable results for standard 4×6-inch prints, you will need a resolution that's higher than 1 megapixel. Therefore, pictures with 0.3-megapixel resolution are not good enough for you to print out, but they appear fairly satisfactory on a computer screen.

Zoom

Optical zoom simply does not exist for most camera phones, due to space and cost limitations. Some cameras have a zoom feature, but it is only digital zoom. Digital zoom cannot increase the quality of the picture. It will only enlarge a portion of the photo at the expense of the resolution. Also, digital zoom will not work if the resolution of your camera is already set at the maximum setting.

Exposure

Usually, little or no control is available for adjusting exposure, apertures, shutter speeds, or focus on camera phones. Some cameras have a low-light or "night-mode" setting, which uses a slower shutter speed to allow extra light to reach the sensor, or a portrait mode with a smaller aperture for shallow depth of field. But these controls are very limited.

Flash

Most camera phones do not have a flash. Nor do they accept a plug for an external flashgun. It is very difficult to take good pictures in low light conditions, and it is more effective to use a separate lamp to light the scene if the lighting level is poor.

General Photo-Taking Tips

Because of the aforementioned constraints, camera phone picture quality is not very high. Here are some general guidelines to follow to take better pictures on your camera phone:

- Take pictures in a well-lit environment.
- Move closer to the subject.
- Put the subject in the middle of the frame.
- Stand still.
- Use high resolution.

Usually after you download a picture to your computer, you still need to use picture-editing software to enhance your photos for optimal results. In the rest of this hack, I will cover some basic image-editing techniques. You can apply these techniques to almost any photo-editing software, including the cross-platform open source GIMP image editor and the Microsoft Photo Editor that comes preinstalled on many Windows computers.

Crop Photos

Nokia camera phones do not have an optical zoom feature. Often, the main subject in your picture will appear quite small because you cannot zoom in when you take the picture. A way to solve this problem is to crop the picture. Figure 11-1 shows the same picture before and after cropping.

Figure 11-1. Cropping a mobile phone photo

By cropping the picture, you enlarge the main subject, achieving a zoom-in effect. You can also eliminate distracting parts (for example, part of another flower in the lower-left corner of the original picture), thereby guiding the viewer's attention to the main subject.

Adjust Brightness and Contrast

A lot of camera phones cannot get exposure timing right. The photo is either over- or underexposed. You can adjust picture brightness and contrast using most picture-editing software. Usually when you are adjusting parameters, you can see how your adjustments affect the picture right away. You can fine-tune the brightness and contrast until the picture looks natural to you. A lot of software programs also allow you to adjust the Gamma parameter. You use the Gamma parameter to adjust a picture's midrange color tones, enabling you to adjust some of the colors that brightness and contrast cannot adjust.

Figure 11-2 shows two pictures before and after adjusting brightness and contrast. In this example, because the picture was taken in a poorly lit environment, I increased the brightness, reduced the contrast, and adjusted the Gamma values.

Figure 11-2. Adjusting the contrast and brightness to enhance a mobile photo taken under poor conditions

Adjust Color Balance and Saturation

Some more powerful picture-editing software (such as Adobe Photoshop CS) also allows you to adjust the color balance and saturation of each color in a photo. To decide whether the general color tone of your picture is balanced, you just need to see if a person's face, or any object that should look white or gray, has the correct color. For instance, the original picture in Figure 11-2 had too much green in it. So, I reduced the level of green and increased the level of magenta. You can do this adjustment for yellow, blue, red, and magenta color tones as well. You can access the color balance tool by selecting the Color Balance feature in your software.

Increasing saturation is another good way to make a picture more vivid. Increasing saturation reduces the shades of gray, black, and white in your photo. You can enhance many pale, dull pictures by using the saturation tool.

Adjust Sharpness

Blurry pictures are a common problem among camera phone photos because of the low quality of the phones' sensors and lenses. So, for many camera phone pictures, it is necessary to use a sharpening procedure to reduce blurriness. Sharpening increases the difference between adjacent gray values in a picture. Figure 11-3 shows the two flower pictures from Figure 11-1 before and after sharpening. You can see that the one on the right is clearer.

—Ju Long

Figure 11-3. Adjusting photo sharpness

Edit Movies on the Phone

Some Nokia Series 60 phones are bundled with software for editing short movie clips. You can then share those edited clips directly from your phone via MMS or some other means.

Some Nokia Series 60 phones, such as the Nokia 7610 and 6630, include two video-editing programs that run on the phone: Vid Editor and Movie Director. You can do some quick video editing with these programs and send the finished movies to friends via MMS or email, without having to turn on your computer.

Vid Editor enables you to combine several video clips in your Gallery into one longer movie. You can select multiple clips and edit them all together in a linear timeline (see Figure 11-4).

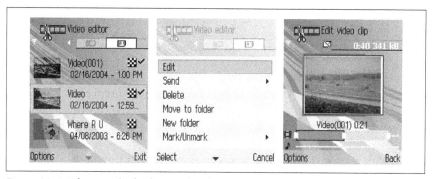

Figure 11-4. Editing multiple clips on the phone using Vid Editor

You can copy and move the clips around along the movie timeline. But you cannot break up or trim each individual clip. The software supports fade-in and fade-out transition effects between clips. It also allows you to apply effects to each clip (see Figure 11-5). For instance, you can turn a color clip into black and white or put it into slow motion. The video editor has a separate soundtrack timeline, where you can import sound clips and add sound effects.

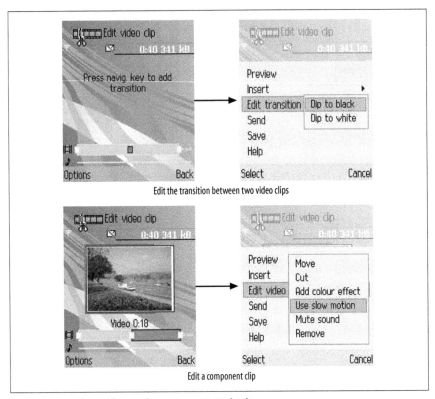

Figure 11-5. Editing clips and transitions in Vid Editor

Movie Director also combines several clips into one movie. But instead of letting you control the effects, Movie Director uses predefined styles to determine the appropriate framing graphics, visual effects, transition effects, onscreen text, and music for the movie. The idea is to spare the novice (or busy) user from worrying about the gory details of movie editing. In quick mode, with a click of a button, you can organize and transform your clips into a fancy movie (see Figure 11-6). In custom editing mode, Movie Director allows you to choose the onscreen text and soundtrack for your movie. Movie Director ships with 10 styles.

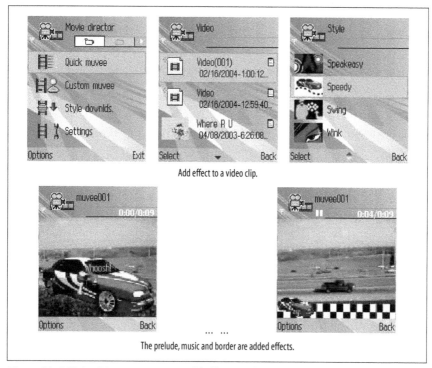

Figure 11-6. *Using Movie Director to add effects to clips*

 ## Share Mobile Movies

#69 Make movies from the phone camera and share them on the Internet for all
to enjoy!

Most Nokia camera phones are capable of recording short video clips. As
such, they are like mini digital camcorders that you can carry anywhere with
you. You can access the video recorder program on a Series 40 phone via the
Media → Camera → Video menu. On a Series 60 phone, the video recorder
program is typically the Video Rec program (e.g., on a Nokia 3650 or 6600
phone) or the Camera → Video program (e.g., on a Nokia 7610 phone)
under the Main menu. A Series 40 phone records video at a resolution of
128×96 pixels, and a Series 60 phone records at 172 ×144 pixels. Nokia
phones typically record video with 16-bit (64k) color at 8–15 frames per second (fps). The video is encoded in the H.263 or MPEG-4 (a.k.a. H.264)
codec and is saved in 3gpp format. You cannot change the color, fps, or
codec settings for your Nokia phone's default video recorder.

> *3gpp* is a video format defined by a consortium of Standards Development Organizations (SDOs) known as the 3rd Generation Partnership Project (3GPP), whose mission is to promote video on mobile phones. To learn more about the 3GPP organization and its work, please visit *http://www.3gpp.org/*. Nokia has adopted the 3gpp standard on all its phones.

A full-color and full-motion video clip, complete with a live audio track, is much more expressive than still pictures. I regularly use my camera phone to capture precious but unexpected moments in life. However, the drawback of video clips is that they take a lot of effort to edit and are difficult to share. Yes, you can send a video clip from one phone to another via MMS. But that is pretty much it. How do you compose and edit several clips into a movie that tells a real story? How do you share the movie across the Internet for everyone to enjoy? I cover all these techniques and tips in this hack.

Convert 3gpp Clips to Other Video Formats

Although it offers superb video compression, the 3gpp video file recorded by the camera phone is not supported on most popular PC video players. The first step to edit and share a video clip is to convert it into a better-supported format, such as the AVI format. You can easily edit an AVI video file or put it on the Internet for everyone to watch. Here I discuss two popular conversion tools.

Use QuickTime Pro. Apple's QuickTime Pro is an excellent video manipulation tool for Mac and Windows computers. You can download QuickTime from *http://www.apple.com/quicktime/*. The free version plays only movies. You need to pay around $30 to buy a Pro license key to unlock the video-editing and saving features.

Then you can transfer the 3gpp video file from the phone to the computer [Hack #33] and open it in QuickTime. Next, choose the File → Export menu to convert it into another format. Near the bottom of the "Save exported file as..." dialog box, you can choose to save it in AVI or MOV format (see Figure 11-7). Click the Options button to set the compression quality, frame rate and other parameters of the converted video file (see Figure 11-8).

Use the 3gpp-to-AVI tool. The 3gpToRawAvi program is a simple and free Windows tool for converting 3gpp video files from Series 60 devices to raw

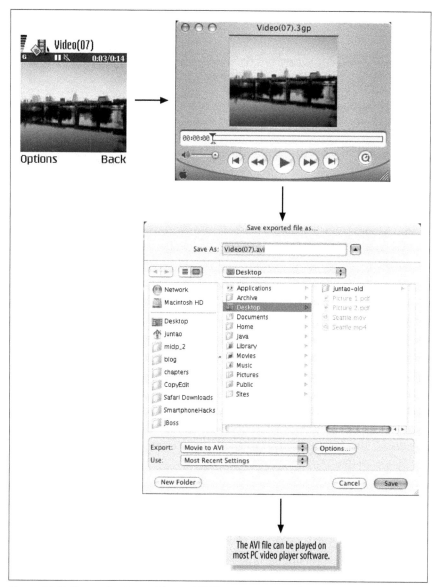

Figure 11-7. Converting 3gpp video to AVI format using QuickTime Pro

AVI files. You can download it from *http://www.allaboutsymbian.com/
downloads/3gpToRawAvi.zip.*

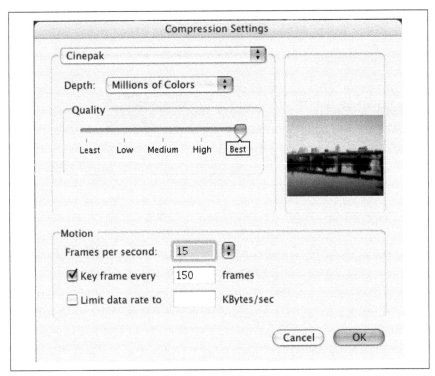

Figure 11-8. Options available when converting the video format

Use the open source FFmpeg tool. If you use Linux or prefer a command-line utility, you can try the open source FFmpeg tool to convert video files. Download the FFmpeg software and documentation from *http://ffmpeg.sourceforge.net/index.php*.

 The FFmpeg program runs on the Windows, Mac OS X, Linux, and BSD operating systems.

Using FFmpeg is very simple. The following command converts the 3gpp-encoded *movie.3gp* file to an AVI-encoded *movie.avi* file:

```
$ ffmpeg -i movie.3gp movie.avi
```

Several other command-line options are available for controlling the frame rate, maximum data flow rate, resolution, color, and many other parameters of a converted video file. For more details, see FFmpeg's documentation.

Edit Movies on the PC

PC video-editing software enables you to make complete movies from the short video clips you take from your mobile phone camera.

Once you've transferred your mobile phone video clips to a desktop computer [Hack #33], you can use powerful video-editing software to make complete and professional-looking movies. Of course, editing mobile phone video is different from editing video taken from a regular digital video camera, since with the former you have to work with short and low-quality clips. In this hack, I'll cover some common video-editing tasks that are most relevant to mobile video editing.

Stitch Together Multiple Clips

The default video recorder in many Nokia phones can capture only 10 seconds or so of video content in each clip. This is to limit the total clip size to less than 100 KB so that it stays under the maximum MMS message size imposed by most networks. The 10-second clips are a bit too short for most events. You have to combine and edit those fragmented scenes before you can share them with friends.

> Some recent Series 60 devices can record longer video clips directly to the MultiMediaCard (MMC) card. For instance, the Nokia 7610 supports 10 minutes and the Nokia 6630 supports up to an hour of continuous recording time.

A second compelling reason to record multiple clips is to maintain acceptable video quality throughout the recording. When using your phone to record video, you should avoid panning too much. Video quality deteriorates when the camera tries to adjust the focus and shuttle speed for different lighting conditions. The best practice is to shoot a fixed subject for a short period of time, turn off the camera, pan the camera, and then shoot the new subject.

QuickTime Pro makes it easy for you to merge clips. For a general example, let's say that you want to append video B to the end of video A. The steps involved are outlined here and in Figure 11-9:

1. First, open the two clips in two player windows side by side.
2. Notice the little triangles below the play bar. In the clip B window, drag the two triangles to highlight the section you want to merge into clip A. You can use the play head to help you determine the start and end frames.

3. The selection is highlighted with gray color on the play bar. Copy the selection to the system buffer via the Edit → Copy menu.

4. In the clip A player, move the play head to the end of the clip and choose the Edit → Add menu to append the buffer content to the end of the current clip.

5. Save the merged clip from the video A player window.

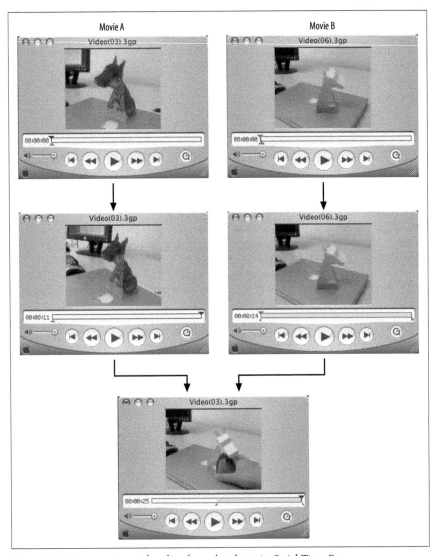

Figure 11-9. Merging two video clips from the phone in QuickTime Pro

Using the two small triangles under the play bar, you can select and copy any section of video from any clip. That allows you to discard any unwanted recordings.

Rotate and Distort Video

Given the low resolution of mobile phone video camcorders, you need to fill the screen with your subject to get as much detail as possible. That sometimes means you have to hold the phone sideways and record in portrait mode as opposed to the default landscape mode. Then you have to rotate the video before you can further edit or share it. The QuickTime Pro software makes it easy to rotate video clips. Just choose the Properties dialog and select Rotate (see Figure 11-10). You can save the rotated clip or merge it with other clips as I discussed in the previous section.

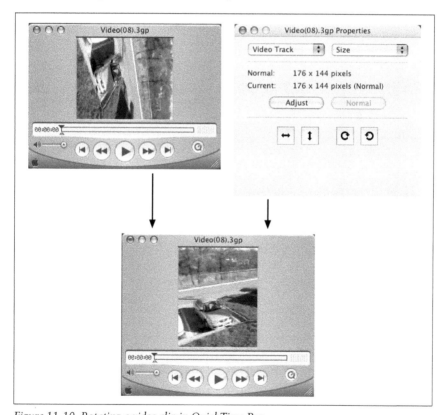

Figure 11-10. Rotating a video clip in QuickTime Pro

QuickTime Pro offers much more functionality than simply rotating a video clip. Figure 11-11 shows that you can rotate video to any angle, and stretch and distort it in any way you want. You can select the Movie → Get Movie Properties menu to bring up the Adjustment dialog box. Click the Adjust button and the player screen shows markers that you can drag to rotate, resize, and distort the video.

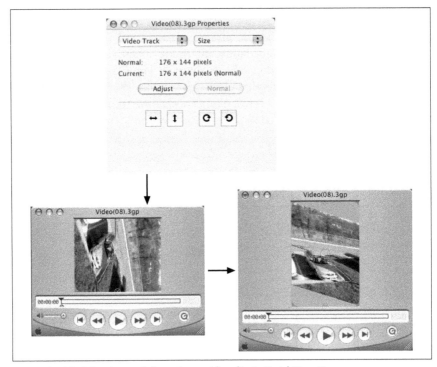

Figure 11-11. Adjusting and distorting a video clip in QuickTime Pro

Edit the Soundtrack

A good soundtrack is key to a successful movie project. However, the recorded soundtrack of mobile video is often not very good due to the limitations of the phone microphone (not to mention the fragmented 10-second clips). QuickTime Pro allows you to remove the original soundtrack, add background music, and even add a narration track to your movies.

To remove the soundtrack from a movie, use the Edit → Extract Track menu and select only the video track in the dialog box. Then save the resulting video-only clip to a new file.

To add a soundtrack, you should first open the audio file in QuickTime. It can be an MP3 music file or your own voice recording. Use the two triangles below the play bar to highlight and then copy a segment of the audio. Place the play head in the video clip to the point where you want to add the sound, and then click Edit → Add to add the soundtrack.

Play PC Video Clips on the Phone

Watch your home video, movie trailer, or even TV programs on your mobile phone. You just need to convert popular PC video formats to a format that can be played on your mobile phone.

Watching video on your mobile phone is cool! Once considered the killer application for broadband wireless networks (e.g., 3G networks), mobile video holds great promise for a whole new mobile lifestyle, with the mobile phone as your entertainment center. If you have a Nokia phone, you can probably watch a lot of video on your device today. Most recent Nokia Series 40 and Series 60 phones, regardless of whether they have cameras, support playback of 3gpp-format video files. The default video player on Nokia Series 60 phones is the Symbian version of RealPlayer.

> An undocumented feature in RealPlayer on Series 60 phones is that you can use the navigation keys (a.k.a. joystick) to fast forward or rewind the movie. Up is fast forward and Down is rewind. You cannot do this with the Nokia video player on Series 40 phones.

Most existing video files are not in the 3gpp format. They need to be resized and re-encoded for mobile phone video players.

Use QuickTime Pro

If you are using a Windows or Mac OS X computer, QuickTime Pro is the best editing tool for video content. You can open any video file and save it as a 3gpp file. Figure 11-12 shows the conversion process.

Clicking the Options button, you can bring up the dialog box for 3gpp movie export. Figure 11-13 shows the video options. You can choose the video size for Series 40 (128×96 pixels) or Series 60 (172×144 pixels) phones. The data rate setting is used to limit the size of the final video file. The low data rate gives you smaller files but also lower video quality (e.g., skipped frames and mosaic effects).

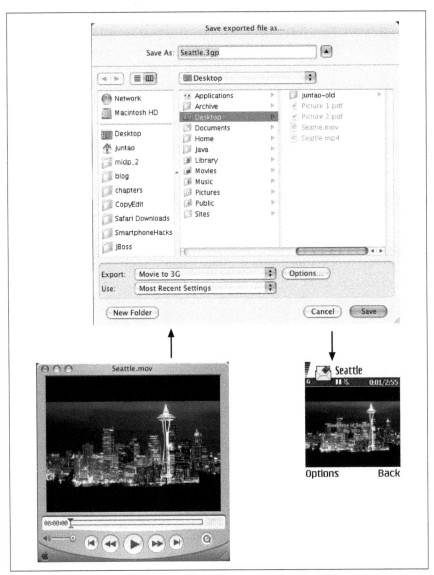

Figure 11-12. Converting a PC/Mac movie file to 3gpp format for phone playback in QuickTime Pro

Figure 11-14 shows the audio options of the exported 3gpp movie. Most Nokia phones can only play mono audio via the phone speaker or headset. If

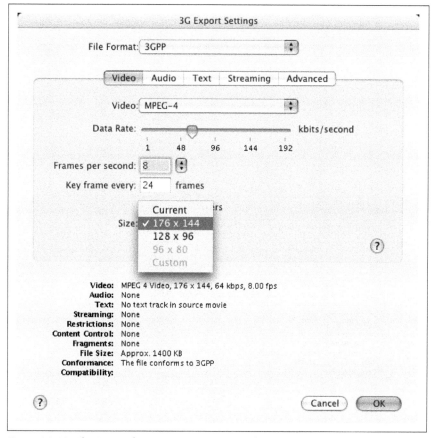

Figure 11-13. The 3gpp video export options in QuickTime Pro

that is the case for your phone, make sure you export the movie audio as "mono" to save space.

For each video file, you have to search for the best compromise of data rates. Table 11-1 lists the data rate settings I typically use. I found them able to produce movies that are of reasonable quality and yet are still small enough to fit into a typical MMC card.

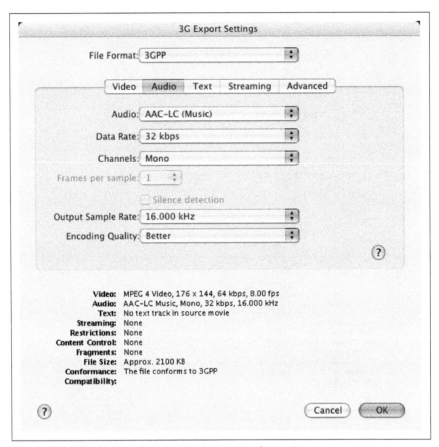

Figure 11-14. The 3gpp audio export options in QuickTime Pro

Table 11-1. Example settings for exported 3gpp movies

Device	Video rate	Audio rate	File size for a two-hour movie
Series 40	24 kbps	12 kbps	32 MB
Series 60	48 kbps	12 kbps	54 MB

 Feature-length movie files are much larger than the 4MB size limit imposed by the phone memory. To transfer movie files to the device via Bluetooth, you need to configure the device message Inbox to save received messages in the MMC card [Hack #21].

The 3gpp standard provides mechanisms for content providers to limit how the user uses the 3gpp movie on the phone (a.k.a. DRM protection controls). For instance, you can forbid the user to forward the movie from the phone, or limit the number of times the movie can be played on the phone. Figure 11-15 shows such options supported by QuickTime Pro.

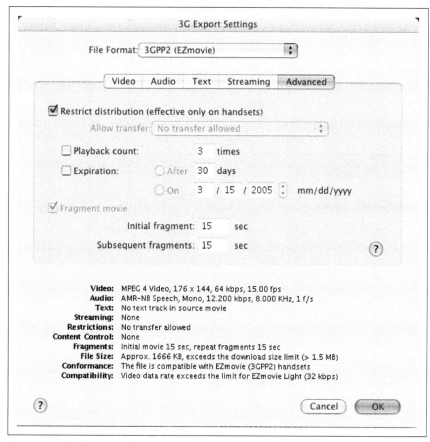

Figure 11-15. The 3gpp advanced options in QuickTime Pro

Use FFmpeg

If you use Linux or prefer a free cross-platform command-line tool, the open source FFmpeg program is best. You can specify the video size, frame rate, data rate, and other configuration options on the command line.

```
$ ffmpeg -i movie.avi -s 172x144 -r 10 \
    -ac 1 -b 48 -ab 12 movie.3gp
```

The preceding command converts the *movie.avi* file to a 3gpp file. The video is resized to 172×144 to match the Series 60 screen. The frame rate is 10 fps. The audio track is set to be mono (-ac 1). The data rates are 48 kbps for video and 12 kbps for audio.

Play DVD Movies on the Phone

Use PC converting software or commercial players to play feature-length DVD movies on your Nokia phone.

In principal, you can use the same basic technique described in "Play PC Video Clips on the Phone" [Hack #71] to convert feature-length DVD movies to the 3gpp format. This produces movies that can be played on both Nokia Series 40 and Series 60 devices. However, you need to go through a few extra steps to make the DVD content ready for 3gpp conversion. In this hack, you will learn exactly what you need to do to play DVD on your phone. In the second half of this hack, I will discuss commercial DVD players for Nokia Series 60 phones. Those commercial tools automate much of the tedious work required to convert DVD movies and support full-screen landscape-mode playback.

Convert DVDs to 3gpp

DVD content is not available in the regular computer video formats, such as AVI or MPEG. Hence, you have to first rip the DVD media files into the hard disk:

- On a Windows PC, you can use DVDx from *http://www.labdv.com/dvdx/*.
- On a Mac OS X computer, you can use the DVDBackup software from *http://www.opuscc.com/download/others.shtml*.

Both programs are freeware. They can remove the regional code encryption on the DVD and copy the DVD content to several *.vob* files in a local directory.

Make sure you own the DVD and are comfortable that you are within your rights to make a backup copy before you proceed.

In a *.vob* file, the video content is encoded in MPEG-2 format, and the audio content is encoded in AC3 format. Unfortunately, QuickTime Pro does not support MPEG-2 out of the box. You have to purchase and install an MPEG-2 playback component before you can save the *.vob* file to a 3gpp file for mobile phone playback.

The FFmpeg program does read *.vob* files as an input source. However, users have noted that the converted 3gpp file often has very low audio volume and distorted pictures. According to a tutorial at *http://excamera.com/articles/24/movie.html*, you can use the MPlayer tool to fix those problems. You can download the MPlayer package for free at *http://www.mplayerhq.hu/ homepage/design7/news.html*. The package includes several command-line utilities that run on all major operating systems. First, use the mencoder command to extract the video track, resize it, and encode it to the MPEG-4 format:

```
$ mencoder movie01.vob -nosound -ovc lavc -lavcopts vcodec=mpeg4 \
    -vop expand=176:144,scale=176:-2 -o movie01.avi
```

The following mplayer command extracts the audio track to a *.wav* file, changes the sampling rate to 8000 Hz, and turns up the volume by 12 db, which is 16 times the power output and sounds about four times louder to the average human ear. The extracted audio file is saved in *audiodump.wav*.

```
$ mplayer -vo null -ao pcm \
        -af resample=8000,volume=+12db:sc movie01.vob
```

Now that you've got separate tracks for video and audio, you can use either QuickTime Pro or FFmpeg to combine and convert the video and audio tracks to a single 3gpp file. The following command listing shows how to use FFmpeg to accomplish this. The -map option specifies which source file is the video track and which is the audio track.

```
$ ffmpeg -i movie01.avi -i audiodump.wav \
        -s 172x144 -r 10 -ac 1 -b 48 -ab 12 \
        -map 0.0 -map 1.0 movie01.3gp
```

> You also can use the MPlayer application to capture video if you have the appropriate video capture or TV tuner card installed and configured. Check MPlayer's reference documentation for more details. Using the technique described, you can, in theory, record and convert TV programs to play on your phone.

Use Commercial Players

Several commercial video player programs are available on Nokia Series 60 devices. Compared with the default 3gpp player, those commercial players offer some advanced playback options and simplify the content conversion process.

SmartMovie. The SmartMovie program from Lonely Cat Games includes a video player for Series 60 phones and a video converter for Windows PCs. You can download and purchase it from *http://www.lonelycatgames.com/ mobile/smartmovie/*. The PC video converter takes any video file supported on your PC (i.e., files that can be played by Windows Media Player on your PC) and converts it to an AVI format the mobile player understands. You can configure the target file's video codec, picture size, and data rates in the converter utility (see Figure 11-16). The converter can also break a long video file into several segments for easy uploading when you have a small MMC card.

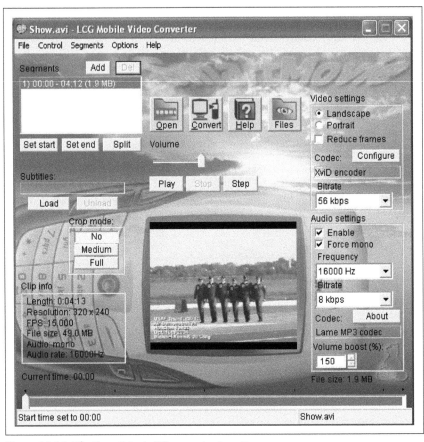

Figure 11-16. The SmartMovie PC converter in action

To play back DVD content on the SmartMovie player, you still need to rip the DVD to *.vob* files and convert them to AVI files. See the description in the previous section on how to do this. The SmartMovie web site also provides a guide for preparing DVD content: *http://www.lonelycatgames.com/mobile/smartmovie/dvd/*.

Once the converted video file is uploaded to the device, you can use the SmartMovie mobile player to play it back. You can play the video in full-screen mode or even in landscape mode (see Figure 11-17) to make the most efficient use of your screen real estate. In landscape mode, the video can be

208×176 pixels, which is almost 45% larger than the regular 3gpp video on Series 60 phones. In addition, the player supports on-screen adjustment of brightness and audio volume, etc., which the default RealPlayer lacks.

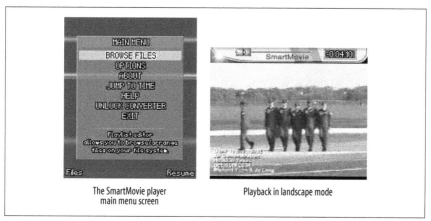

The SmartMovie player
main menu screen

Playback in landscape mode

Figure 11-17. The SmartMovie player playing AVI movies in full-screen landscape mode on the phone

DVDPlayer. Despite its name, the DVDPlayer from Viking Informatics does not actually play DVD disks or *.vob* files on your phone. However, it does play AVI files encoded with MPEG-4 Xvid or DivX codecs. Those AVI files are commonly found in P2P networks for movies or TV shows. You need to resize them to 208×176 pixels and set a reasonable limit for data rates, as I discussed earlier in this hack. The DVDPlayer plays the movie in full-screen landscape mode. You can download and purchase the software from the vendor's web site, *http://www.vikinggames.hu/product.php?id=23*. This site also has a comprehensive tutorial on how to rip and prepare DVDs for mobile phone playback.

Monitor the Real World

HACK #73

Use the phone as a baby monitor, a webcam, a home intrusion detector, and even an obstacle detector for the blind.

With their camera and sound recorder, our mobile phones can extend our eyes and ears to places we cannot reach. Compared with a regular voice recorder or web camera, the mobile phone is always connected to the network, and hence is always accessible. In this hack, I'll discuss some innovative software solutions that turn your phone into a multipurpose audio- and video-monitoring device.

Baby Monitor

Psiloc's Baby Care program turns your Nokia Series 60 phone to an audio monitor for infants. You can place the phone in the baby's bedroom and activate the program. The phone monitors the sound level in the room. When the baby wakes up or makes noise, the phone will automatically call or send a Short Message Service (SMS) message to a specific number to alert you so that you can come and check on the baby.

You can also configure the Baby Care program to accept phone calls from a specific number (the "parent number"). When you call the phone from that number, the phone will not ring, and instead, just silently accepts the call. This way, you can hear the baby and speak to the baby through the speaker phone whenever you want.

Figure 11-18 shows the configuration options in Baby Care, including the sound detection sensitivity, the parent number, the notification options, and whether to silently receive calls from the parent. You need to keep the Baby Care program running in the background to keep it functional.

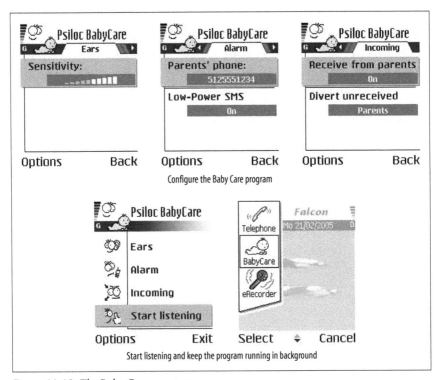

Figure 11-18. The Baby Care program

Of course, you also can use the Baby Care program as a spying device. You can leave a phone running the Baby Care program somewhere, and then call in to listen to all the nearby activity. Use it for such purposes at your own risk!

Security Camera

Extending the idea of an audio monitor, you can use your camera phone as a video monitor. Together with motion-detection software, you can make the phone call a security company or send out a picture message via MMS or email whenever it detects motion.

The Nokia Observation Camera is a wall-mounted security camera with a General Packet Radio Service (GPRS) communications module built in. You can configure it to take pictures at a preset interval and send the stream of images to a web server (i.e., like a regular webcam). It can also be triggered by motion or a change in temperature.

Webcams typically feed live images to an Internet server for viewing in web browsers. With special software such as Eye-SpyFX (*http://www.eyespyfx.com/mobileviewing.asp*), you can view webcam images anywhere, at any time, on your mobile phone as well.

At the time of this writing, no publicly available mobile phone–based security camera solutions are available. But I am aware of several such projects that are underway. So, watch out for the news, and we might soon see one of them in the marketplace!

An Eye for the Blind

Another innovative use of the phone camera is as an obstacle detector for the blind. The vOICe BEB program is a free Java application for camera phones. You can download it from *http://www.seeingwithsound.com/voice.htm*. The vOICe BEB program takes snapshots from the camera at fixed intervals and then the software scans through the pictures horizontally to turn them into short audio clips. The tone at each time corresponds to the pixel levels and distributions on each scan line. According to the information on the web site, at any scan line along the image width, the image height is associated with pitch and the image brightness is associated with volume. For example, a bright, rising image on a dark background sounds as a rising pitch sweep, and a small bright spot sounds as a short beep.

Although vOICe is a Java program, it does not yet work on Nokia Series 40 devices, since those devices do not support access to the camera via Java applications.

Figure 11-19 shows the vOICe program in action. You need to give it permission to access the camera [Hack #16].

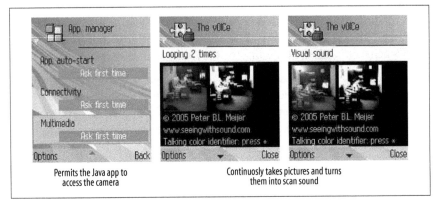

Permits the Java app to
access the camera

Continuosly takes pictures and turns
them into scan sound

Figure 11-19. The vOICe program in action

For the user to use this program, she must first train herself to understand the meaning of the sounds. PC-based programs are available that can help train users to do this. You can learn more about these PC programs and about this technology from the vOICe web site.

Refer to "A User Interface for the Vision Impaired" [Hack #48] to check out options for enhancing the mobile phone user interface for the elderly and vision impaired.

HACK #74 Play Digital Music

Turn your phone into an MP3 music player.

From the Sony Walkman to the Apple iPod, personal music players are among the most successful mobile devices in tech history. As the mobile phone becomes the convergence point of personal communications and entertainment, we all want it to double as a music player so that we no longer need to carry multiple devices.

Most Nokia phones have the capability to play digital music files:

- Almost all Nokia phones can play Musical Instrument Digital Interface (MIDI) files. MIDI files contain musical notes, and the phone emulates musical instruments to play those notes. The advantage of MIDI files is that they are extremely small. A typical MIDI file is only several kilobytes. You can store thousands of them in your phone's memory. The disadvantage is that MIDI files can only simulate musical instruments, not voices. So, you cannot have any actual songs recorded in MIDI.

- All Series 60 phones and many Series 40 phones can play WAV files. WAV files store a range of spectrums in the sound wave, so, they can record any sound the human ear can hear. The disadvantage is that WAV files are typically very large. A complete five-minute song can take tens of megabytes of memory space. So, you can store only song segments on the device.

- Some devices, such as the Nokia 3300 in Series 40 and Nokia N-Gage in Series 60, have the capability to play MP3 files. MP3 files contain heavily compressed sound data. The compression algorithm is optimized for the hearing characteristics of the human ear. MP3 files are much smaller (about 5 MB for a five-minute song) than WAV files, with equivalent sound quality. It is the ideal format for personal music players. You should check your phone manual to see if it supports the MP3 format. For Series 60 devices, you can install third-party software programs to play MP3 files (described later in this hack).

Although we have yet to see a hard disk–based "iPod phone" from Nokia, you can purchase a large MMC card to store music. For instance, a 512MB MMC card can store hours of MP3 music. This is comparable to the capability of many flash-based dedicated MP3 players in the market. So, I focus on MP3 playback in this hack. First I will discuss how to obtain MP3 content and then I'll discuss third-party MP3 player programs on Series 60 devices.

 Most Nokia phones also have radio receivers for AM/FM programming. You must have an earphone plugged in (or a connected Bluetooth headset) to listen to the radio.

Get MP3s

You can download MP3 files from the Internet or rip them directly from audio CDs you own. Many software programs can help you rip audio CDs, including WinAmp and MusicMatch for Windows, XMMS for Linux, and iTunes for Mac OS X. It is important to note that most Nokia devices can play music only in mono mode, even if the music file contains stereo data. That is a limitation imposed by the phone hardware. So, when you prepare an MP3 for the phone, you can generate files in mono format to reduce the file size. A key parameter that determines the MP3 file size is the bit rate. Higher-bit-rate MP3 files have larger sizes but also produce better sound quality. On Nokia phones, typically a 32kbps bit rate is sufficient.

Then you can use the Music Transfer application in the Nokia PC Suite to upload a set of MP3 files from your PC to your phone. Of course, you can also transfer files one by one using many other means [Hack #33].

> The Nokia Music Transfer application in the Nokia PC Suite also allows you to rip MP3 files from an audio CD. This way, you can transfer the entire contents of an audio CD to your phone in one step.

Play MP3 Music

If your phone supports MP3 playback, you can start to enjoy the MP3 music in your Gallery right away. However, what if your phone does not support MP3 playback out of the box? If you have a Series 60 phone, you can install a third-party MP3 player program to play back the MP3 files. In this section, I'll discuss two such programs.

> In fact, even for devices that have built-in support for MP3 files, the software players provide some cool features such as MP3 metadata display, rich player control, and playlist management.

Mp3Player. Mp3Player from Viking Informatics plays MP3 files on your phone and displays information about the songs using metadata stored in the MP3 file headers. Its playlist includes all MP3 files on the device. Mp3Player is capable of playing music in stereo mode on devices that have stereo hardware support. You need to check your device's manual to see if your device supports stereo audio. You can download and purchase Mp3Player from *http://www.vikinggames.hu/product.php?id=1*.

UltraMP3. The UltraMP3 player from Lonely Cat Games plays not only MP3 files, but also a variety of other music formats, such as MOD, XM, IT, and S3M. The MOD (module) format is especially interesting. It contains both MIDI-like notes for the music and WAV-like audio for voice. As a result, module music files are very small and are of limited quality compared with

MP3 files. The module format was widely used by music bands as a "preview" format for their songs. The relatively low sound quality of module music files is not a big concern for the limited mobile phone speakers.

Another important feature of the UltraMP3 player is that it provides a playlist editor. Playlists allow you to control exactly how your songs are played. They are essential for improving the user experience and managing music files, especially when you have a large collection. The UltraMP3 player also supports custom skins. You can customize it with graphics that fit your own style. Figure 11-20 shows the UltraMP3 playback controls and the playlist in its standard skin.

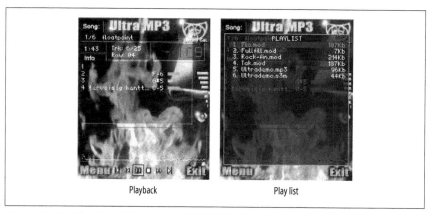

Figure 11-20. The UltraMP3 player in action

 ## Record Audio
#75
Use your phone as a digital voice recorder.

Almost all Nokia phones come with a built-in voice recorder. You can use it to record short sound clips into WAV or Audio/Modem Riser (AMR) format audio files in the Gallery for later playback or sharing. It is an excellent tool for capturing sound bites in everyday life. You can also use the recorder to capture ring tones [Hack #45].

The voice recorder is also used to record voice commands. Each command is a short audio clip that is associated with a specific action on the phone, such as launching an application or dialing a phone number. For instance, you can record the word *camera* as a voice command for launching the Camera application. After the command is recorded, whenever you say the word "camera" to the phone, it matches the recorded sound wave and launches the Camera application.

However, a drawback of Nokia's built-in voice recorder is that it records only 60 seconds of audio. With large MMC cards available, the phone can store much longer audio clips, which opens new opportunities for cool applications. In this hack, I'll cover third-party audio software programs for Series 60 devices.

Record a Long Audio Clip

The Extended Recorder (i.e., eRecorder) program from Psiloc (download and purchase it from *http://www.psiloc.com/index.html?id=167*) lets you record long audio clips. Figure 11-21 shows that it can record hundreds of minutes of audio in several MB of memory space. The Extended Recorder also provides extensive playback controls.

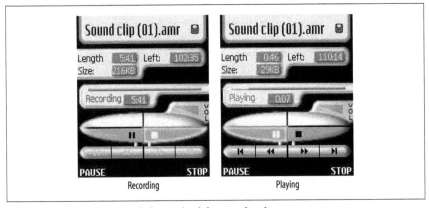

Figure 11-21. Recording and playing back long audio clips

In fact, your recording length is limited only by your available memory or the size of your MMC card. Extended Recorder essentially turns your phone into a digital substitute for a tape-based portable voice recorder. It is excellent for recording interviews, classes, and conferences. Or, you can use it to capture your thoughts and ideas while on the go. The recorded WAV file can be easily transferred to computers [Hack #33] for backup, compression (e.g., to MP3 format), or even voice-to-text conversion.

Control Recording at Any Time

If Extended Recorder is running in the background, you can use predefined hotkeys to start and stop recording while the phone is in idle mode, running another program, or even in the middle of a phone conversation. Figure 11-22 shows how to set the recording hotkey to the Pen/ABC key +

the green Call key. It also shows the start and stop points of the recording when the hotkey combination is pressed in the phone's idle mode.

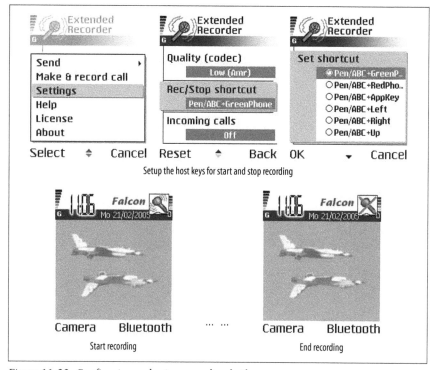

Figure 11-22. Configuring and using recording hotkeys

You also can use Extended Recorder to record phone conversations. See "Record a Phone Call" [Hack #31] for details.

Index

We'd like to hear your suggestions for improving our indexes. Send email to *index@oreilly.com*.

H

Handango web site, xix, 75, 81
hardware for Bluetooth, 58
hciconfig tool, 66
hcid daemon, 67
hcitool tool, 66
Headset Profile, 55
Hilton, Paris, 123
Hollowman calculator, 30
Holtmann, Marcel, 65
hotkeys
    controlling recording with, 358
    defined by Opera browser, 236
Howard Forums web site, xix, 12
.htaccess file, 259, 260
HTML (Hyper Text Markup
        Language), 232
    configuring servers for MIME
        types, 257–260
    displayed by Services browser, 234
    email client, sending/receiving
        messages via, 297–299
    making phone calls from web
        pages, 263
    mobile web sites, creating, 253
    Opera browser, 234–237
    WebViewer browser, 240
Hugo, Dominique, 94

I

ICQ (IM system), 319
iDEN (Integrated Digital Enhanced
        Network) networks, 3
idle screen
    changing fonts on, 223
    customizing, 216–222
    operator logo, changing, 221
idle-screen soft keys, assigning
        applications to, 90
IM (see instant messaging)
Image Converter program (Nokia PC
        Suite), 219
image sensors in camera phones, 329
images as wallpaper
    from camera phones, 220
    from phone Gallery, 217
images, turning off to speed up Opera
        browser, 238

IMEI (International Mobile Equipment
        Identity) number
    blacklist database, adding to, 121
    locating, 20
    proxy servers identifying browsers
        via, 239
    registering Symbian applications, 83
IMPS (Mobile Instant Messaging and
        Presence Services), 326
incoming calls
    receiving, during active
        conversations, 138
    recording, 141
information theft, protecting
        against, 117
Infrared Data port, beaming files
        via, 154, 166
instant messaging (IM), 319–327
    Agile Messenger program, 322–323
    future of, on phones, 325–327
    interacting with, via mobile web
        portals, 249
    via portals, 324
    via SMS, 321
interactive_console.py script, 87
Intercom Profile, 55
international formatting for phone
        numbers, 131
Internet
    choosing data plans, 14–20
    connecting laptops to, via
        tethering, 14, 16
    connecting Linux computers to, via
        mobile phone
        modems, 184–189
    connecting Macs to, via mobile
        phone modems, 182–184
    connecting PCs to, via mobile phone
        modems, 181
    connecting smartphones to, 43–51
    sharing movies from camera
        phones, 335–338
    tethering laptops to, 14, 16, 180
Internet Connect program, 183
Internet Information Services (IIS),
        associating MIME types with
        files/directories, 259
Internet Relay Chat (IRC), 319
    clients available for mobile
        phones, 320

iPhoto, launching from phone with
    Salling Clicker, 199
IR (infrared)
    beaming files via Infrared Data
        port, 154, 166
    connecting computers to wireless
        networks, for SMS
        messages, 309
iSync program, 169–172
iTunes, controlling music with Salling
    Clicker, 197–199

J

J2ME
    aggregating and rendering RSS
        content on phones, 271
    TipicME IM client, 324
    writing applications using, 79
Jabber, 319
JAD (Java application descriptor)
    files, 75–77
JAR (Java archive) files, 75–77
Java applications
    configuring, after installing memory
        card, 100
    installing from computers, 76
    installing over the air, 75
    managing, 78
    running on smartphones, 74–80
    writing your own, 79
Java-based web client
        (WebViewer), 240
Jepson, Brian, xii, 20, 272
JIC DCT4 Unlocker, 34

K

Kapadia, Kamil, xiii, 131
KDE Bluetooth Framework, 68
    downloading, 154
kernel requirements for Bluetooth
        support under Linux, 64
keyboards, external, using with Series 60
        devices, 282
Keypad tones, factory setting for, 207
keypads, typing on, 281–286
Koehntopp, Frank, xiii, 327

L

l2ping command, 67
LAN Access Profile, 55
laptops, connecting to Internet via
        tethering, 14, 16
launching applications quickly, 91
Laurie, Adam, 107
LCD backlight brightness, adjusting, 37
leaking out personal information, using
        Bluetooth scanners, 105
lenses in camera phones, 329
Lindholm, Christian, x
Linux
    configuring Bluetooth for, 64–69
    connecting computers to Internet via
        mobile phone
        modems, 184–189
    using Bluetooth GUI tools with, 154
    using gnokii to transfer data between
        PCs and phones, 163
Linux Unwired, 64, 189
Litefeeds service, 271
LiveJournal (blog provider), 275
local business listings, searching Google
        via SMS, 316
local devices, examining, 66
locked phones, unlocking, 28–34
locking
    memory cards, 119
    phones, 117
Locknut virus, 105
Log program on Series 60 devices, 23
    reducing size of log, 101
logo, creating your own on the idle
        screen, 221
Lonely Cat Games, 297, 349, 356
Long, Ju, xiii, 146, 332
long-distance calling cards, 136
lost phones, 116–122
    reporting, 121
loudspeaker, putting phone on, 139
low battery message, 37
Lundberg, R. Emory, xii, 14, 36, 131,
        178, 286

M

Mac computers, running serial consoles on, 88
Mac Mail program, launching from Salling Clicker, 202
Mac OS X
  Bluetooth File Exchange program, 151
  configuring Bluetooth for, 62–63
  connecting
    computers to wireless networks, for SMS messages, 309
    to Internet via mobile phone modems, 182–184
  open source J2ME toolkit for, 80
  synchronizing phone data with, 169–172
  using phones as remote controls for, 196–205
Mac OS X Panther Hacks, 51
Mac system, controlling, using Salling Clicker, 203
Mack, Darla, 226
magnifying parts of screen (for the vision impaired), 224
mailboxes, setting up, 292
maintenance of batteries, 41
malicious programs, 104–111
  basics of, 106
  preventing, 108
  removing viruses, 109
MANGOobjects, themes developed by, 225
manual backup and restore, 115–116
MapQuest directions, sending to mobile phones, 177
masters/slaves (piconets), 54
MBus interface cables, working with unlocking software, 34
media files, using FExplorer with, 95
media library, browsing, using remote control, 194
media playback, controlling PCs for, 190–195
Meeting profile, 207
memory cards
  backing up main memory to, 114
  installing, 99
  locking, 119
  protecting data on, 119

memory compression software, 101
memory management techniques for smartphones, 99–103
MercuryXMS software package, 311
Merlin (GSM/GPRS PC card), 309
Message alert, factory setting for, 207
message templates at Yahoo! IM portal, 325
Messaging application
  accessing received files, 96
  configuring, after installing memory card, 100
  on Series 60 devices, 287, 292–297
metered data plans, 14
  comparison of, 18
  switching out of, 19
microphone, limitations for recording soundtracks, 342
Microsoft Messenger (MSN), 319
MIDI (Musical Instrument Digital Interface) songs
  playing on phones, 355
  using as ring tones, 214
Midlet Review, 75
MIDlets, 74
  filesystem locations for, 97–98
MightyPhone service (FusionOne), 174
MIME (Multipurpose Internet Mail Extensions) types, configuring servers for, 257–260
#MIN# (#646#) service number
  T-Mobile, 25, 34
  Verizon Wireless, 36
*MIN# (*646#) service number (Cingular), 35
miniGPS program, 211
Minute Man National Historical Park, audio tour of, 146
MinuteCheck and account usage alerts, 26
Minutes Manager (Moov Software), 23
minutes offered by plans, 13
minutes-tracking software for Series 60 devices, 23
MMC cards
  backing up main memory to, 114
  deep resets and, 113
  exchanging files between phones/PCs, 155
  installing, 99

N

National Mobile Phone Register
    (UK), 122
NetWireNews program, launching from
    Salling Clicker, 202
networking technologies for
    phones, 2–4
    choosing the right one, 9–14
networks, Bluetooth, 53–55
New York, audio tour of, 145
news, delivering to phones, 314
news sites, combined in portals by
    Moreover, 251
Nextel
    interfaces to SMS, 290
    metered data plan from, 18
    unlimited data plan from, 18
nights and weekends, free minutes
    on, 13
Nokia 3650 camera phone, 8
    Nokia PC Suite and, 70
Nokia 5140 phone, 6
Nokia 6170 phone, 6
Nokia 6230 phone, 6
    defining access points, 49
Nokia 6600 phone, 222
Nokia 6600/6620 camera phones, 8
    Nokia PC Suite and, 70
Nokia 6820 phone, 6
Nokia 7210 phone, 6
Nokia 7610 phone, 222
Nokia 7710 phone, 9
Nokia 9300 Communicator, 8
Nokia 9500 Communicator, 8
Nokia Application Installer program, 71
Nokia Audio Manager program, 71
Nokia browser, 233
Nokia Connection Manager
    program, 72
Nokia Contacts Editor program, 72
Nokia Content Copier program, 71
Nokia Developer's Suite for J2ME, 79
Nokia Image Converter program, 71
Nokia Mobile Internet Toolkit
    (NMIT), 261
Nokia Modem Options program, 72
Nokia Multimedia Converter
    program, 216
Nokia Multimedia Player program, 72

Nokia N-Gage phone, 8
    exchanging files between
        phones/PCs, 155
Nokia Observation Camera, 353
Nokia PC Suite, 69–73
    automatic backup and restore, 114
    backing up/restoring PIM data, 116
    connecting
        computers to mobile
            networks, 181
        computers to wireless networks,
            for SMS messages, 309
        to devices, 72
    downloading/installing, 69
    exchanging files between
        phones/PCs, 148
    Image Converter program, 219
    installing JAD and JAR files, 76
    installing .sis files on devices, 82
    limitations of, 72
    manual backup and restore, 115
    programs in v6.4, 71
    remote data, protecting, 122–124
    synchronizing phone data with
        PCs, 168
    transferring contacts from
        computers, 158
Nokia PC Sync program, 71
Nokia Phone Browser program, 71
Nokia Sensor application, 53, 166
Nokia Series 40 Theme Studio, 226–228
Nokia Series 60 Theme Studio (for
    Symbian OS), 226–228
Nokia Sound Converter program, 72
Nokia Text Message Editor program, 72
Nokia WAP Gateway Simulator, 262
Nokia Wireless Keyboard, 282
Nokia WML Browser Emulator, 262
NokiaFREE calculator, 30
NokiaFree Forums web site, xix, 226
nonrechargeable batteries, using as
    emergency backup
    solution, 40
normal resets, 112
.nth files, 228
#NUM# (#686#) service number
    (T-Mobile), 34
*NUM (*686) service number
    (T-Mobile), 35

Colophon

Our look is the result of reader comments, our own experimentation, and feedback from distribution channels. Distinctive covers complement our distinctive approach to technical topics, breathing personality and life into potentially dry subjects.

The image on the cover of *Nokia Smartphone Hacks* is tin cans and string. A tin can telephone is probably one of the great joys of youth. It is fairly easy to construct one. It works through the basic physics of sound. When the speaker holds his end of the phone—his can—in front of his mouth and speaks into the can's open end, the sound waves travel into the can, causing the can to vibrate, which in turn causes the connecting string to vibrate. And so the sound waves travel along the string to the other can. The second can vibrates, and the sound comes out of the can. Simple, yet perfect!

Mary Brady was the production editor for *Nokia Smartphone Hacks*. Audrey Doyle was the copyeditor. Mary Brady proofread the book. Philip Dangler and Claire Cloutier provided quality control. Judy Hoer wrote the index.

Hanna Dyer designed the cover of this book, based on a series design by Edie Freedman. The cover image is an original photograph by Getty Images. Karen Montgomery produced the cover layout with Adobe InDesign CS using Adobe's Helvetica Neue and ITC Garamond fonts.

David Futato designed the interior layout. This book was converted by Keith Fahlgren to FrameMaker 5.5.6 with a format conversion tool created by Erik Ray, Jason McIntosh, Neil Walls, and Mike Sierra that uses Perl and XML technologies. The text font is Linotype Birka; the heading font is Adobe Helvetica Neue Condensed; and the code font is LucasFont's TheSans Mono Condensed. The illustrations that appear in the book were produced by Robert Romano, Jessamyn Read, and Lesley Borash using Macromedia FreeHand MX and Adobe Photoshop CS. This colophon was written by Mary Brady.

Have it your way.

Get even more for your money.

Join the O'Reilly Community, and register the O'Reilly books you own. It's free, and you'll get:

- $4.99 ebook upgrade offer
- 40% upgrade offer on O'Reilly print books
- Membership discounts on books and events
- Free lifetime updates to ebooks and videos
- Multiple ebook formats, DRM FREE
- Participation in the O'Reilly community
- Newsletters
- Account management
- 100% Satisfaction Guarantee

Signing up is easy:

1. Go to: oreilly.com/go/register
2. Create an O'Reilly login.
3. Provide your address.
4. Register your books.

Note: English-language books only

To order books online:
oreilly.com/store

For questions about products or an order:
orders@oreilly.com

To sign up to get topic-specific email announcements and/or news about upcoming books, conferences, special offers, and new technologies:
elists@oreilly.com

For technical questions about book content:
booktech@oreilly.com

To submit new book proposals to our editors:
proposals@oreilly.com

O'Reilly books are available in multiple DRM-free ebook formats. For more information:
oreilly.com/ebooks